Cambridge Imperial and Post-Colonial Studies Series
General Editors: Megan Vaughan, King's College, Cambridge and Richard Drayton, Corpus Christi College, Cambridge

This informative series covers the broad span of modern imperial history while also exploring the recent developments in former colonial states where residues of empire can still be found. The books provide in-depth examinations of empires as competing and complementary power structures encouraging the reader to reconsider their understanding of international and world history during recent centuries.

Cambridge Imperial and Post-Colonial Studies Series
Series Standing Order ISBN 978–0–333–91908–8 (Hardback)
978–0–333–91909–5 (Paperback)
(outside North America only)

You can receive future titles in this series as they are published by placing a standing order. Please contact your bookseller or, in case of difficulty, write to us at the address below with your name and address, the title of the series and the ISBN quoted above.

Customer Services Department, Macmillan Distribution Ltd, Houndmills, Basingstoke, Hampshire RG21 6XS, England

---

The Bengal Delta

Ecology, State and Social Change, 1840–1943

Iftekhar Iqbal
Assistant Professor of History, University of Dhaka, Bangladesh

First published 2010 by
PALGRAVE MACMILLAN

Palgrave Macmillan in the UK is an imprint of Macmillan Publishers Limited, registered in England, company number 785998, of Houndmills, Basingstoke, Hampshire RG21 6XS.

Palgrave Macmillan in the US is a division of St Martin's Press LLC, 175 Fifth Avenue, New York, NY 10010.

Palgrave Macmillan is the global academic imprint of the above companies and has companies and representatives throughout the world.

Palgrave® and Macmillan® are registered trademarks in the United States, the United Kingdom, Europe and other countries.

ISBN 978–0–230–23183–2   hardback

This book is printed on paper suitable for recycling and made from fully managed and sustained forest sources. Logging, pulping and manufacturing processes are expected to conform to the environmental regulations of the country of origin.

A catalogue record for this book is available from the British Library.

A catalog record for this book is available from the Library of Congress.

10   9   8   7   6   5   4   3   2   1
19   18   17   16   15   14   13   12   11   10

Printed and bound in Great Britain by
CPI Antony Rowe, Chippenham and Eastbourne

For Rizwana

Contents

List of Figures and Tables

Figures

Tables

Preface and Acknowledgements

This book seeks to situate ecology within the matrix of long-term social and economic change in the colonial Bengal Delta. Focused on a period between approximately 1840 and 1943, it examines the place of ecology in the changing relationships between the colonial state and different social and political forces. The terms 'East Bengal' 'eastern Bengal' and 'Bengal Delta' are used interchangeably, denoting the region that corresponds to most of today's plain lands in Bangladesh. 'Ecology' and 'environment' are used interchangeably, although environmentalists see subtle difference in the scope and meaning of the two terms. The official spelling of some places has changed over time, notably Dhaka (from Dacca), Kolkata (from Calcutta), Barisal (from Backergunge or Bakarganj), Sundarbans (from Soonderbunds or Sunderbans) and Comilla (from Tippera or Tipperah). I have generally used the recently standardized names in the text.

The book was conceived during my doctoral research in Cambridge where Christopher Bayly's supervision was vital. It was a joy working with him and I remain indebted to his guidance, critique and encouragement that have made this project viable. David Arnold, Rajnarayan Chandavarkar (who would have been delighted to see this book, but for his untimely departure), Joya Chatterji, David Ludden, Willem van Schendel and Jon Wilson read the full text or parts of the book-in-progress. Their interventions were crucial. Conversations with the following colleagues on different occasions proved immensely valuable as the book was taking shape: Mumtaz Ahmad, Daud Ali, Peter Bertocci, Gautam Bhadra, Ranjan Chakrabarty, Gunnel Cederlöf, Vinita Damodaran, Rajat Datta, Richard Drayton, Rohan D'Souza, Richard M. Eaton, Richard Grove, Tim Harper, Mark Harrison, Ira M. Lapidus, David Lewis, Michael Mann, Peter Marshall, John McGuire, Iris Möller, Ralph W. Nicholas, Joseph O'Connell, Mahesh Rangarajan, Peter Robb, Francis Robinson, Dietmar Rothermund, Tirthankar Roy and Chris Smout. During my exciting research trip to Kolkata, Binay Chaudhuri, Amalendu De, Chittabrata Palit and Rajat Ray generously shared their thoughts and helped me to navigate through relevant source materials. Md Touhid Hossain, the then Deputy High Commissioner of Bangladesh to India (later Foreign Secretary to the Government of Bangladesh), showed great interest in the project and provided logistic support in accessing government institutions in West Bengal. Our time in Kolkata was made most enjoyable because of the cordial hospitality of Zuglul Kabir and his family.

In Bangladesh I was enormously benefited from conversations with my teachers and colleagues including Sharif Uddin Ahmed, Fakrul Alam, Nurul H. Choudhury, Abdul Momin Chowdhury, Syed Anwar Hussain, Harun-or-Rashid, Iftikhar-ul-Awwal, Mufakharul Islam, Muntasir Mamun, Tazeen Murshid, Kazi Shahidullah, Sirajul Islam, Ahmed Kamal, Muinuddin Ahmad Khan, K.M. Mohsin, Nurul Huda Abul Monsur, Asha Islam Nayeem, Mahbubur Rahman, Badruddin Umar and Shireen Hasan Usmani. I am particularly indebted to Sharif Uddin Ahmed for his most needed support and guidance at the National Archives of Bangladesh where he was the Director.

Among fellow colleagues and friends, my sincere thanks are to Sunil Amrith, Rachel Berger, Ai Lin Chua, Neilesh Bose, Nandini Chatterjee, Ashfaque Hossain, Isaraf Hossain, Ananya Kabir, Mike Lewis, Kris Manjapra, Tilottama Mukherjee, Ian Petri, Emma Reisz, Haimanti Roy, Jayeeta Sharma, Jon Wilson and Rashed Zaman. Jon means a lot for the book. Over more than a decade, we spent hours in Jessore, Dhaka, Cambridge, London and on the skype to discuss the project for which he has been a key critic and inspiration. Life in Cambridge was all the more enjoyable because of friends like Mushtaq Ahmed, Zhimming Cai, Fakhruzzaman Fakhru, Shahed Imam, Raihana Islam, Sadia Khan, Saleh Naqib, Adam Lauridsen, Fatima Marzia, Premjith Sadasivan, Libby Sang and John Stubbs. I am also greatly indebted to Mahtab U. Ahmad, Bill Allison at Fitz, Nicola Cavaleri, Sue Free, Sheikh Abdul Mabud and Sandra Welsh for their valuable support on different occasions.

A rich range of local source materials for the book is drawn from the archives in Bangladesh, the UK and India. Useful excavations of these resources were possible because of the generous support of concerned staff. I wish to register my thanks to the staff of the following archives and libraries: in Bangladesh: National Archives of Bangladesh; Dhaka University Central Library; Bangla Academy Library; Varendra Research Museum library in Rajshahi; Libraries and Record Rooms of Comilla, Faridpur, Jessore, Rajshahi and Chittagong. In India: National Library of India, West Bengal State Archives, and the Writer's Building in Kolkata. In the UK: the British Library, especially the Asia, Pacific and Africa collections; University Library, Seeley Library of History Faculty, Library of Faculty of Oriental Studies, Central Science Library and Centre of South Asian Studies Library in Cambridge; Maugham Library of King's College London, SOAS Library, and Wellcome Library in London. Special thanks are to Pranab Kumar Chattapadhyay (the then Secretary, West Bengal State Archives), Kevin Greenbank (Centre of South Asian Studies, Cambridge), Mazharul Islam (National Library of India, Alipore) and William Noblett (Official Publications wing, University Library Cambridge).

The research for the book would have been impossible without the Cambridge Scholarship for Commonwealth Studies from Cambridge Commonwealth Trust. The Smuts Memorial Fund provided financial support for conducting fieldwork in India and Bangladesh. Scholarships, awards and grants from Fitzwilliam College, Faculty of History and the Board of Graduate Studies at Cambridge substantially contributed to my academic well-being. I am indebted to the Trustees of the All Saints Educational Trust, Leche Trust and Charles Wallace Bangladesh Trust for their generous support. Grants from the Royal Historical Society and Economic History Society enabled me to travel to and retrieve materials from the British Library. At the final stage of writing-up, a substantial scholarship from St Luke's Institute provided the most needed peace of mind, and I am indebted to David Lake for this reason. The book finally sees the day of light because of the encouragement, practical support and patience of Richard Drayton, Michael Strang, Ruth Ireland and Barbara Slater at Palgrave Macmillan.

Some of the material in this book has appeared in journal articles and I would like to make the following acknowledgements: Cambridge University Press for 'Return of the Bhadralok: Ecology and Agrarian Relations in Eastern Bengal, c. 1905–1947', *Modern Asian Studies*, 43(6) (2009); White Horse Press for 'Fighting with a Weed: Water Hyacinth and the State in Colonial Bengal, c.1910–1947', *Environment and History*, 15(1) (2009); and the Arnold Bergstraesser Institute for 'The Railways and the Water Regime of the Eastern Bengal Delta c.1845–1943', *Internationales Asienforum: International Quarterly for Asian Studies*, 38 (3–4) (2007).

My parents, Khondker Sarwarul Islam and Ummul Hasnat Siddiqua, are happiest to see the book coming through. Their constant moral support has been a guiding force whether at home or away. My greatest debt is to Rizwana Siddiqua. Her encouragement and passion for this project remained constant at a time when she alone had to take care of the nitty-gritty of domesticity and our son Ridwan, who grew up with this book.

II
Dhaka, March 2010

List of Abbreviations

ABP	Amrita Bazar Patrika
ABR	Assam-Bengal Railways
Add MS	Additional Manuscripts
APAC	Asia, Pacific and Africa Collections, British Library, London
BLCP	Bengal Legislative Council Proceedings
BLAP	Bengal Legislative Assembly Proceedings
BoR	Board of Revenue
BPBEC	Report of the Bengal Provincial Banking Enquiry Committee
BS	Bengali shon/year
BTA	Bengal Tenancy Act
CBI	Communication, Building and Irrigation Department
CCRI	Dept of Co-operative Credit and Rural Indebtedness
CSAS	Centre of South Asian Studies, Cambridge, UK
Comm	Commissioner
Dept	Department
Div	Division
EBR	Eastern Bengal Railways
EBDG	Eastern Bengal District Gazetteers.
GoB	Government of Bengal
GoI	Government of India
IJMR	*Indian Journal of Medical Research*
IOR	India Office Records, British Library, London
LRC	Report of the Land Revenue Commission
Misc colln.	Miscellaneous collection
MoE	Minutes of Evidence, in RIC
NAB	National Archives of Bangladesh, Dhaka

Offg	Officiating
Progs	Proceedings
RIC	Report of the Indigo Commission
RAB	Report on the Administration of Bengal
Rev	Revenue
RWCSB	Report of the Working of the Cooperative Societies of Bengal
RWHC	Report of the Water Hyacinth Committee, Bengal
Secy	Secretary
TBCJ	*The Bengal Co-operative Journal*
WBSA	West Bengal State Archives

Glossary

abad	cultivation, mostly in the reclaimed lands from forests and alluvial formations
abadkar	one who clears jungle and cultivate
abhadra	non-bhadralok
abwabs	unauthorized extra taxes
adivasi	indigenous people/forest dwellers or hillsmen
amalnama	writ giving preliminary possession
anna	sixteenth part of a taka or rupee
apadkala	time of natural disaster or misfortune
ashraf	Muslim aristocrats, claim to have their origin in the Middle East
atmashakti	self-prowess, confidence
atraf	ordinary, non-ashraf people
aus, aush	a kind of paddy harvested in August–September
bahubal	inner strength of mind
barga	system of sharecropping
baor	low lands running like water channels in jungle areas; used also for sheets of water ordinarily cut off from rivers
bastu	homestead land
banyan	a local manager for a European merchant or East India Company servant
bapari	agent of business concerns
Barais	a dealer in betel nuts
bhadra	gentlefolk
bhadralok	see note 23 of chapter 1, p. 196
bhag	shares; produce rent

bhati	downstream direction, deltaic plains of East Bengal
bid'a	sinful innovation (in Islam)
bigha	about one-third of an acre
beel, bil	marsh, discarded beds of rivers; uncultivable land producing grasses only or low cultivable land growing 'boro' variety of paddy
char, chur	alluvial accretion; land formation off fluvial actions of deltaic rivers
chitha/chitta	paper showing measurement of fields
dai	traditional midwife
dalan	masonry building
daula bari	second home
dastur	old customs
dhan	paddy
diara	alluvial accretion/formation
doob	tender grass overgrowing new chars
dowl	settlement
faquir, fakir	saint, beggar
goalas	cowherd, milkmen
goor	molasses
gomastha	village agent of indigo concern
hahakar	loud lamentation; hopelessness
hat	scheduled weekly or biweekly rural market
hawala	literarily 'a charge', a kind of tenure
hawaladar	one who takes tenures to clear land and settle raiyats, mostly in the Sundarbans and in the *diara-chars*.
haor	kind of marsh
hookah-pani-bandh	social boycott
hundi	money brokering
ijara	lease
itor	non-bhadralok

jamabandi	rent/revenue roll
jihad	holy war
jot, jote	land, lot; may constitute a tenure or holding
jotedars	richer peasants/village oligarchs
jow	state of soil when it is perfectly moist for a plough after the early showers of spring
jumma bandi	rent roll
jumma, jamma	rent ceiling
jummah	Friday prayer of the Muslims
kabuliat	counterpart of lease or settlement contract
khal	natural or artifical water bodies, usually narrower than a small river
khalifah	local leader of Faraizi circle
khas	One's own, private; also the property of government (e.g., khas mahal)
khas mahals	estates owned by government
khas tahsil	collections from khas mahals
khedda	a system of catching wild elephants
khodkasht	resident, especially refers to raiyat
kismet	fate
lakh	one hundred thousand
lascar	sailor/seaman
lathial	professional clubmen
madrasa	Islamic traditional religious educational institution
mahajan	merchant, money lender
mahal, mehal	patches of land, estate
maulovis	traditional scholars of Islam
maund	40 seers (82 lbs avdp)
mofussil, mofossal	suburban area, small town
mouza	a revenue unit corresponding to a village or a number of villages
mukhtear	legal agent/attorney

munshi	a secretary or assistant
munsif	civil judge in a local court
nazr, nuzzur	gift, bribe
nirick	rate
nolok	nose ring
nullah	kind of small stream, canal
paikasht	non-resident, especially referring to raiyat
panchayet	a traditional rural institution of reconciliation and justice, composed of local elderly and influential people
phandidar	police constable
pice	one fourth of an anna
pir	Muslim saint
pottah, patta	written leases, title deeds
puja	Hindu religious festival, worship of gods and goddesses
pukur	tank or pond
puthi	popular Bengali poetic narratives
rasadi	progressive rent
ratha yatra	Hindu festival in which chariots are employed in the procession
raiyat, ryot	Persian word for peasant, cultivator
riwaj	traditional Muslim religious practices
roab	goodwill
ryatwari	system of settlement of land directly between the government and the tenant
salami	bonus, gift
seer	2 lbs avdp
shirk	idolatry
swadeshi	early 20th-century nationalist movement in Bengal favouring home-industry and boycott of foreign goods
Swaraj	self-governance, home-rule

taluqdar/talukdar	a holder of small estate, often a ceded portion of a zamindari
tasbih	beads
thana	police station
touzi, tauzi	unit of land measurement
tola	toll, cesses
tehsil, tahsil	a revenue circle
tahsildar, tahsildar	native collector of rent
uparasta khalifa	superior representative of the Faraizi movement
ustad	teacher; a Faraizi leader
veeta	raised earth on which rural homestead are built
zamindar	landlord
zamindari	authority and jurisdiction of a zamindar
zenana	women's quarters

1
Introduction

This book is a study of political practice and social and economic change taking place in a dynamic and mutually constitutive relationship with environment. Environmental history has taken great strides in the past few decades, in the global as well as the South Asian context, yet, apart from peripheral regions, such as forests and hills, ecological questions remain absent from the broader history of South Asia and many other parts of the world. With a long-term historical focus on the Bengal Delta, which approximates to today's Bangladesh, this book argues that an understanding of the ecology of plains is essential for any analysis of the politics and society of colonial South Asia. (Figure 1.1 shows Bengal and neighbouring territories; Figures 1.2 and 1.3 show the geological setting of the Bengal Delta and the Himalayas and the water regime that defines the region.)

A nuanced and long-term perspective on the environmental history of the Bengal Delta should take into account the spatial specificities of the ecological regime of north-eastern South Asia: high mountain peaks, the deltaic plains and the vast coastline of the Bay of Bengal. A key feature of the landscape of the region is the combined river system of the Ganges, Brahmaputra and Meghna rivers, together constituting the largest delta on earth. These rivers, winding in innumerable branches through the delta and emptying into the Bay of Bengal, carry the highest proportion of sediment of any world river system, amounting to about 25 per cent of the world total. It is estimated that some 40,000 million cubic feet of silt settle in the deltaic plain on their journey to the Bay of Bengal, creating enormous areas of new land, known as chars and diaras.[1] The process of sedimentation reaches its peak during the annual monsoon downpour. At the same time, the ocean currents are impeded by the heavy outflow from the rivers, and in turn deposit a

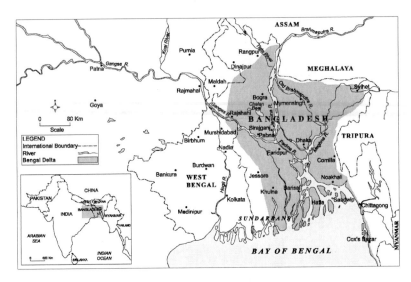

Figure 1.1 Map of Bengal and neighbouring territories

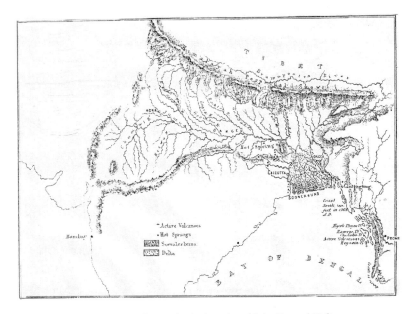

Figure 1.2 Map showing the geological setting of the Bengal Delta
Source: *The Calcutta Review*, March 1859.

Figure 1.3 Map showing the Himalayas and the water regime of the Bengal Delta
Source: J.D. Hooker, *Himalayan Journals: or, Notes of a Naturalist in Bengal, the Sikkim and Nepal Himalayas, the Khasia Mountains, &c* (London, 1854).

huge amount of sand in the coastal region. Thus a double process of land-making continues amidst the mutual confrontation of silt-laden rivers and the sand-carrying sea.[2] This process of land formation has been encouraged by various species of mangrove, which have facilitated the consolidation of the shoreline through natural succession.[3] Maps of the region drawn in AD 1548, 1615, 1660 and 1779 by Gastaldi, de Barros, van den Broecke and James Rennell respectively, including those drawn throughout the colonial period, testify to this continuous process of land formation.[4] The process of land formation appears to have been followed by the growth of different vegetation and plants which turned into forest if left uncleared. In this way the Sundarbans emerged between the Bay of Bengal and the fringes of the Bengal Delta.

At the turn of the nineteenth century, the Sundarbans covered an area of about 10,000–12,000 square miles.

In the nineteenth century, most of this gradual increase in land was taking place in the eastern half of Bengal or what comprises most of postcolonial Bangladesh. This was mainly the result of the gradual eastward movement of the Ganges, beginning in the sixteenth or seventeenth century, from the Hugli-Bhagirathi channel towards the bed of a smaller stream, called the Padma. The eastward shift of the Ganges ended when it met the river Brahmaputra near Dhaka in the early nineteenth century.[5] One theory is that this eastward shift of the Ganges was caused by the gradual rise of the level of the land in north-western Bengal and the subsidence of the Bengal basin together with a gradual eastward tilting of its overlying crust.[6] Another theory, dating from 1861, was that the Brahmaputra and other large rivers in the delta rose and fell about a month sooner than the Ganges, with the result that when the floods in the Ganges were at their highest, the Brahmaputra had already started to fall. This provided 'a natural vent for the Ganges to the south-east', whilst all overflow to the south-west was 'gorged'.[7] This shift of the Ganges dealt a severe blow to the formerly flourishing region of western Bengal. The situation was further aggravated by a number of major earthquakes and inundations in the late eighteenth century, which displaced the rivers of northern Bengal while eastern Bengal continued to be ecologically alive and 'active'.[8]

What historical sense can we make of this highly fluid and sea-faced frontier territory of Bengal characterized by an extensive span of silty Sundarbans and alluvial land-forms? Richard Eaton's authoritative analysis of the social history of this agro-ecological zone covers the long medieval period, leaving the colonial period for separate treatment.[9] In an attempt to examine the changing relationship between ecology and society in colonial times, this book studies three broad issues: the process of change in the region's ecological regime itself; the ways in which changes affected the rural economy and society; and the pattern and impact of the ways in which ecological resources were appropriated by different social forces over a longer span of time. In charting the changing relationship between Bengal's fluid and forested ecology and the well-being and poverty of the population, the book argues that the deterioration of East Bengal's ecology, in particular the decay of river and forest systems, which was largely the result of complex political interventions, intensified communal antagonisms which powerfully influenced the political course of the region in the twentieth century as well as the regional economy and society.

Although the dynamic nature of the ecological regime of eastern Bengal makes it a potentially important case study in environmental history, it has failed to attract the attention of environmental historians. Despite remarkable achievements in the environmental history of South Asia, the field largely remains concerned with the realm of nature and its relations with the state. There are signs that the environmental history of South Asia, after initially dealing with the normative discourse of 'destruction' and 'conservation' of forestry in the colonial period,[10] is diversifying thematically. But the question of how to locate nature within broader political and social practices in colonial times remains largely unaddressed, particularly in the context of the plains of South Asia. Another reason for the attraction of environmental history to peripheral landscape is because it emerged within Indian scholarly debates when subaltern studies were at their height. Environmental history took up the unique opportunity to situate the question of power and resistance in the context of ecological regimes in which these actors, such as adivasis, forest dwellers and hillsmen, lived and operated. In the process, however, larger domains of plain land, such as the Bengal Delta, have been left out of environmental history.[11]

While such lacunae in South Asian environmental history have yet to be seriously addressed, some recent interventions have sought to bridge the gap between the 'agrarian' and the 'environmental'. Mahesh Rangarajan suggests that 'the "ecological" half of environmental history would only make sense if it is tied with the broader enterprise of the new social history'.[12] In other words, without the merging of the ecological and the social, the 'agrarian' ceases to be a meaningful category. Such engagement may form the foreground of what Agrawal and Sivaramakrishnan call the 'third generation' of environmental history. Yet, the way in which a merger between the agrarian and the environmental is advocated may not be enough to enable us to grasp the dynamics of the agro-environmental history of the colonial Bengal Delta. The present call for the study of the 'agrarian environment' seems to consider the categories of 'environment' and 'agrarian' as corresponding so harmoniously that other categories, such as non-peasant or urban social forces and their agrarian-environmental practices and imagination, are left out. Such a conceptualization of exclusive harmony between the 'agrarian' and the 'environmental' can prove detrimental to the immense potential of environmental history for South Asia, as exemplified in a recent otherwise important work on South Asian environmental history.[13]

If the environmental historians of South Asia have at least become aware of the importance of connecting to agrarian history, the mainstream histories have altogether failed to situate ecology. The political, cultural and economic stories of the plain lands have been told and retold in a framework in which environment has either no place or has been used as 'geographical background' – a fixed ecological bow from which the arrows of all kinds of history take flight, the ecology itself remaining ontologically static. The best that we have seen recently is the connection of ecology to demographic trends or a Tagorian romantic appreciation of the riverine tract of Bengal. That ecology is central to – or at least a major catalyst of – mainstream political and social histories remains unappreciated. Before we proceed further it is, therefore, important that we take a quick overview of mainstream, especially agrarian, historiographical debates on Bengal.

II

At the beginning was the question of the Permanent Settlement, the land revenue system introduced in 1793, which created a class of zamindars (landlords) bestowed with almost absolute power over land and the collection of revenues on behalf of the colonial state.[14] Issues relating to the Permanent Settlement's ideological origin, complex tenural systems and agrarian institutional arrangements, the relationship that it helped to define between the landlord and the peasants – all have been examined for clues to the stagnation of the rural economy and society in the colonial period.[15] Criticism of the Permanent Settlement grew increasingly acute in the course of the nineteenth and twentieth centuries, and the first major attempt to break with this element of the colonial legacy was made with the abolition of the system by the governments of East Bengal (Pakistan) and West Bengal (India) in the 1950s. An intriguing question arises here: if the structural mechanisms of colonialism were held to be the key to the ills of agrarian society and rural economy, why did abolition of the settlement fail to bring about the anticipated mobility and 'development' in the six decades following decolonization?

Such questioning of the formal, elitist and 'structural' analysis of colonial agrarian society necessitated a shift of focus towards the 'return of the peasant'[16] in South Asian historiography, yielding a vast literature that deserves a little detailed treatment. In the particular case of East

Bengal, the restoration of 'agency' to the largely Muslim peasantry can be examined in three broad perspectives: 'patron-clientelist', 'world-capitalist' and 'subalternist'. An important version of the 'patron-client' analysis of peasant society was elucidated in the 1970s by Ratnalekha Ray, who argued that a section of the richer peasants or village oligarchy, identified by her as jotedars, had been powerful catalysts in agrarian relations as early as the Mughal period. They sustained patron-client relationships and socially reproduced and exercised power over the vast majority of the subordinate peasantry by extending credit and market facilities. These groups remained key social and political forces even after the creation of the zamindars through the Permanent Settlement.[17] More recent versions of the patron-client thesis suggest that richer peasants, instead of investing in capitalist farming, perpetuated the system's grip on the vast rural mass as creditors, traders and rentiers, which resulted in rural stagnation.[18]

The 'world-capitalist' analysis of Bengal agrarian society is comprehensively employed in the works of Sugata Bose. Examining links between the globally connected capitalist system and conditions in local agrarian production, Bose sought the source of the peasant's dominant role within the context of the world economic depression of the 1930s. Although Bose believes that the power of the zamindars declined as a result of a series of tenancy acts legislated since the 1880s, he argues that during the economic depression of the 1930s richer peasants were able fully to assert their influence by extending credit to the vast numbers of poor peasants, creating a situation of extreme dependency and vulnerability that culminated in their being the main victims of the great Bengal famine of 1943. Thus the rich peasants became the 'chief beneficiaries' of the system, followed by zamindars and grain-dealers in the wake of the famine. Bose points out:

> the peasant elite which had begun to separate itself out from the bulk of smallholders in the 1930s further strengthened its position during the catastrophic decade of war, famine and partition. As the relative price of jute remained weak and the older categories of mahajans refused to return to the money-lending business, east Bengal peasants suffered heavily in the great famine of 1943 and accounted for a disproportionate share of the massive quantities of total land alienated.[19]

In other words, the emergence of the rich peasant in agrarian eastern Bengal, a process Sugata Bose calls 'kulakization', coincided with soaring poverty that culminated in the great Bengal famine of 1943, leaving the rich peasants well-placed to further expand their base of domination by buying up the holdings of famine victims.[20] Within the subaltern studies project the case of East Bengal peasant society was taken up by Partha Chatterjee. Chatterjee does not disagree with Bose's conclusion about the emergence of rich peasants as 'surplus appropriators' in the course of the first half of the twentieth century, but he has strong reservation about Bose's methodological approaches. He criticizes Bose for analysing the dominance of the peasant from an angle that results 'entirely from a reified structural dynamic' of demography, market and credit relations and not from the 'conflict among conscious human agents'.[21] The idea of consciousness among East Bengal peasant society had already been employed in Chatterjee's important essay in the first volume of *Subaltern Studies*:

> the notion of community continued to act as a live force in the consciousness of the peasantry, which still treated feudal landlords or agents of the state as outside claimants on their obedience and their produce ... there were differences in the way a peasant community would view itself vis-à-vis landlords who were clearly outsiders as opposed to those who had risen from the peasantry to become part of the rentier class.[22]

As obvious victims of 'false consciousness' peasants preferred to ally with their dominant co-religionists rather than to be dominated by the Hindus, who were considered 'external'. This position did not help the ordinary peasant to escape the suffering of material deprivation. Communal, rather than class, consciousness not only perpetuated the hegemony of the richer peasants but lent support for the Muslim communitarian idea of Pakistan which promised a 'utopia' that would free its inhabitants from the domination of 'outsiders' such as the Hindu zamindars or bhadralok[23] and mahajans. The twentieth-century East Bengal peasant therefore ended up supporting a communally inspired nation-state rather than organizing a more materialistic 'peasant revolution'.[24]

Thus, on the wreckage of an older historiography that dealt with the zamindars and bhadralok as catalysts for agrarian relations, the peasant has been revived, albeit that what started as the 'return of the peasant' ended as a spectre of the same. The thrust of this peasant-centric approach is that the peasant has become more influential than the zamindars in the analysis of complex agrarian relations. After a slight earlier twirl over whether these domineering peasants should be called 'jotedars' or 'rich peasants', the latter prevailed and the only difference between the historians endorsing the category of the peasantry centred on their chronology of peasant dominance. Ray believed that influential richer peasants had been present in both pre-colonial and colonial times, Bose found them active and powerful since the 1930s and Chatterjee found the rich peasant's dominance in an immeasurable realm of consciousness which seemed to transcend time.

The problems with this historiography are as remarkable as its career. It emerged not only in response to the lopsided attention given to colonial institutional structures and elites, but as a critique of the way Indian nationalism was represented as a vertically-placed and detached high politics by the so-called Cambridge School of Indian history in the 1960s and 1970s.[25] This new historiography attempted to prove that Indian nationalism was indeed horizontal and was close to national aspirations of liberation and well-being. In the process, as far as Bengal was concerned, the new historiography largely lost sight of the inner cleavages of nationalist politics. The implication of the renewed attempts to lend broader legitimacy to Indian nationalism has been that the return of the peasant is occasioned by the retreat of the 'zamindari bhadralok'. As Bose argues, in the wake of the Depression the rentier and trading classes, including Hindu talukdar and mahajans, 'ceased to perform any useful function'.[26] For Partha Chatterjee, although the bhadraloks' 'principal income was no longer from rent', they could afford this as many of them had 'white-collar employment, or trade or the professions'.[27]

The emphasis on the economic emergence of the dominant peasant, alongside the decline of agrarian society and the gradual detachment of the previously predominant zamindar-bhadralok and mahajans from the agrarian landscape, lends a strangely romantic attitude to the patriarchy of the good old days, but fails to answer some of the questions that the new historiography itself raises. If sections of the peasantry were potentially capable of exerting their influence over the poorer peasants from the late nineteenth century, why did they wait until the depression of the 1930s to take the upper hand? How was it possible for the richer peasants to strengthen their exploitative actions as creditors and buyers of the lands of poorer peasants at a time when relatively wealthier zamindar or bhadralok failed to do so in the context of the economic depression of the 1930s which affected all – particularly all strata of the peasants – who must have suffered from the drastic fall of prices? Why didn't communalism become more intense in the nineteenth century when the Hindu zamindari bhadraloks were supposedly more assertive? Why did communalism become so fervent at a time when the Hindu bhadraloks' authority and power had supposedly declined in the countryside? The new turn in peasant studies applies the subalternist idea of autonomy to an understanding of resistance in the specific context of East Bengal. The subalternist ethos is liberating enough to explore each of the autonomous categories: of adivasi, of caste, of Muslim peasant or bhadralok elite. This process helps the subalternist historians to add deeper meaning to the ideas and activities of each of these autonomous categories. Yet, because the categories of community, class and power are so hidden behind the thick walls of autonomy, the scope for the reciprocal interactions of historical agencies is somewhat restricted. In the process, as David Ludden puts it, 'all the histories of all the empires and nations in South Asia could never capture the history of all its peoples'.[28] One way to proceed on the question of locating power in the spatial and temporal realities of social activities, as Sivaramakrishnan argues, is to 'examine the every day forms of power, to describe the ambivalences, contradictions, tragedies, and ironies that attend it, and do so in particular locations, situating subaltern and elite in regional histories'.[29]

This book is an attempt to rescue the 'agrarian' from its 'autonomous' cell. True, the concept of autonomy has been challenged and such critiques seem to have informed more recent subalternist scholarship on the different regions of India.[30] Yet East Bengal peasant society remains under its conspicuous shadow.[31] I will argue that the peasant and the bhadralok, the countryside and the towns, power and powerlessness, class and consciousness, politics and culture, well-being and poverty are so intractably entwined that historians need to rework the genealogy of the postcolonial historiography of Bengal. This book takes an ecological perspective that is particularly encouraged by the recent critique of the idea of autonomy that seeks the 'fetishization of social difference', translating it into the difference between 'agrarian' and 'environmental', among others. The relative failure of the agrarian history of modern Bengal in placing itself in broader, multiple political and social contexts can be understood in terms of the way in which it has remained disassociated from the ecological questions.[32]

III

Most literature on the environmental history of South Asia tends to see the state as the key authority and manager of nature.[33] For historians who believe that conservation practices started during colonial times, the idea of a powerful state is crucial – it managed the botanic gardens, patronized scientists and ran a 'colonial laboratory'.[34] Those who argue for a 'watershed' theory of unprecedented destruction of nature during the colonial period also proceed on the assumption that the state, with its hunger for accumulation of profit, rent-seeking and development planning, demonstrated its indivisible political authority. The subaltern resistance that takes place within specific ecological spaces in the wake of both the conservation and destruction efforts of the state are also taken as an index of the state's power against which such resistance is launched. In the past three decades C.A. Bayly's intervention has altered the received notion of colonial power. In examining the flexible and formative adaptability of Hindu merchants or Muslim servicemen Bayly has shown that the state may not be the only functioning location of power.[35] More recently Jon Wilson has shown how discourses of colonial authority and reform emerged from the incomplete nature of colonial power. It was the authority's attempts to cope with the uncertainty of everyday life in an unknown world that had such a tremendous transformative effect on rural Bengal.[36] Given these revisions, where do we place the ecological regime of East Bengal within the competing discourses and practices of power?

K. Sivaramakrishnan has recently argued that in the face of relative intractability and illegibility in the frontier forest tract of south-western Bengal, known as Jungal Mahal, the East India Company acknowledged the limits of its authority among certain tribal groups, prompting an 'enduring pattern of exceptionalism in the political administration' of the colonial state. Until the British were able to form, as Sivaramakrishnan argues, authoritative knowledge about the region through cadastral survey and plot-by-plot demarcation and classification in the early twentieth century, there had existed what he calls 'zones of anomaly' within the imperial standard of administration. This is exemplified by the fact that the Permanent Settlement was not applied in these woodland spaces, implying a significantly different relation between the state and social production. Sivaramakrishnan argues: 'after the Santhal hool (uprising) of 1855 this form of rule grew more sophisticated, and in addition to the suspension of civil and criminal law as normally applied elsewhere … [these non-regulation provinces] also had special tenancy law that sought to prevent land alienation from tribal indigenous to non-tribals'.[37]

Gunnel Cederlöf criticizes Sivaramakrishnan's idea of 'anomaly' for its suggestion that things were 'normal' in non-forest areas.[38] However, as far as eastern Bengal is concerned, I would argue that the region offered a broader space for 'anomaly' until about the turn of the twentieth century when the colonial state began to be more knowledgeable about the interior of the countryside. An uncomfortable fall in state income, caused by the decline of Bengal indigenous industries and the introduction of the Permanent Settlement as well as growing criticism of the company's role in the occurrence of the famine of 1770 and subsequent deterioration in economic and social well-being, forced the East India Company to take a fresh look at the highly fluid chars and forests of eastern Bengal for the purposes both of generating revenue and mitigating peasant unrest. By the 1830s the reclamation drive was vigorously under way.[39] As the depression of the 1830s subsided, paving the way for a rise in price of agricultural commodities, there began an era in which the unique ecological domain of the delta became doubly significant – for the Company, private investors and jobless commercial craftsmen, and crucially for landless peasants.[40] The reclamation of forests and chars for agriculture became part of a general worldwide process in the mid-nineteenth century, which was regarded as the great age of commodity demand and consequent change in land uses.[41] This was also a time when the British colonies, after a long spell of closed mercantilist practice, were increasingly becoming open to the wider world.[42]

As we shall see in the book there were many areas, particularly reclaimable forests and chars, where the revenue code of the Permanent Settlement was not introduced at all or where it was practised with calculated indifference. For the colonial state, revenue was the first claim and because the Permanent Settlement fixed the revenue ceiling after the 1790s, any possibility of alternative revenue sources was liberally exploited. The first target of the 'anomalous' policy of the state was the landlord, against whose interest the peasants were trusted with reclamation and production processes which demanded hard physical labour and ecological skills appropriate to the fluid landscape. Though revenue was the grand motif of the state, the rent rate itself had to be lenient. Along with this, surplus production and commercialization of jute and rice, the enormous water transport network, and old and newly emerging bazaars inaugurated a degree of buoyancy that was unseen elsewhere, with the exception of a few places, such as Lower Burma.[43] There is enough evidence to infer that peasant society in this period was comparatively less stratified, hugely productive for domestic consumption and the market, less exposed to communal violence, and showing signs of relative well-being. The political resurgence of the peasant in the nineteenth century was expressed by the Faraizi movement, which represented the power of the peasant through a movement that was organized, efficient and long-lasting – a scenario that historians have attributed both to the movement's adherence to an agro-Islamic world-view and to the idea of secular class struggle.[44] But the world-view that the Faraizis represented was neither modern nor nationalist. It remained an agrarian one. It was born and flourished in the heart of the agro-ecological frontier of Bengal and declined – coincidentally – in the wake of the emergence and strengthening of the nationalist movement around the turn of the twentieth century, which was also a time when poverty and ecological decline were beginning to surface.

Whether we consider the nineteenth-century social and economic dynamism of East Bengal as a result of an 'anomaly' within standard colonial rule or whether we see it more generally as a picture of continuity from the Mughal period, the issue of the power of the state remains unresolved in the longer perspective of colonial history. It gives the impression that the 'modern' state is restrained and less assertive when it is less knowledgeable about the agro-ecology of a certain region, but that it begins to assert its power when it acquires the appropriate understanding. By attributing the assertiveness of the agency of the indigenous peasant society to the 'ignorance' of the state, the capacity of the peasantry to generate and employ its own power – that is different to the power of the state – is overlooked. The idea of 'anomaly' also pays limited attention to the question of what happens to the agrarian social and economic relations when the period of 'anomaly' ends. Do the material and political conditions of the social actors undergo change? Do they continue to receive the patronage of the state or do new agencies evolve at the interface of the colonial state and the local social and political forces? What happens when the state is more knowledgeable about nature, when it is in a position to classify and codify the ways in which ecology and society inform each other? Some important work in this area, including that of James Scott, is perceptive of the power of the state, but does not attempt to examine the social forces that make this power of the state possible.[45] Beyond the concept of anomaly we must, therefore, also examine the discursive register of power as both subject and object, within and outside the structural logic of the state. In other words, it may be argued that in the nationalist phase of colonial history, indigenous agency operated in its own way even where the state was fully knowledgeable – the state's power only ever acted by working on the powers of others that best suited it.

The possibility of such developments have eluded the major historians of South Asia largely because of the ambiguous and contradictory messages coming from the discourses of twentieth-century anti-colonial nationalism, which considered the institutions of modernity – states, bureaucracies, newspapers, universities and academic disciplines – 'to be external and foreign to the natural ecology of Indian agrarian environments'. 'But, in fact', David Ludden argues, 'moderns invented traditional agrarian environments endowed with all the elements of culture, nature, technology, and religion.'[46] In this view, the early twentieth-century anxiety over the decadence of the Hindu nation, the idea of the end of 'Sonar Bangla' (Golden Bengal) under colonial rule and a patriotic urge to return to the village as part of the Swadeshi movement all seem to be connected by the unbroken thread of the power that was to be. Thus beside focusing on the specific category of the 'colonial', 'agrarian' or 'environmental' a more useful intervention would be to take a deeper look at the category of the 'national' in locating the power that engaged in agrarian relations within the ecological regime of the Bengal Delta.

IV

Once we have secured an ecological perspective of the social construction of power and its changing relations to the state and nation, questions regarding the other two interrelated issues, that is, changes in the ecological regime itself and their corresponding impact on well-being and poverty, can be comprehended better. Environmental problems in East Bengal have long been attributed to coastal cyclones and flooding. However, while life and property at the coastal fringe were always vulnerable to cyclones, the inhabitants of the region seem to have sustained techniques specifically adapted to the productive use of flooding. It was the annual monsoon flooding that prepared the ground for the following year's schedule of cultivation and production. Until the late nineteenth century, flooding continued the age-old process of renewing surface soil with silt, and flushing the majority of the territories with fresh water, which had the effect of lowering mortality rates due to disease in comparison with moribund regions of West Bengal. However, this formative agro-ecological domain began to change remarkably around the turn of the twentieth century following the arrival of the railways in East Bengal. The railways began to operate in the western part of Bengal in the late 1860s but spread to the interior of East Bengal by the 1890s, the networks becoming a significant feature of the region's landscape by the 1930s. Being built almost entirely on embankments criss-crossing the vast deltaic plains, the railway divided the delta into 'innumerable compartments'.[47] The result was a total disruption of the free-flowing water systems and loss of the concomitant benefits of the spread of silt and salubrious flushing of the landscape. The Ganges-Brahmaputra water system, which sustained a remarkably productive agricultural regime for over a millennium, came to face its most formidable man-made obstruction. These problems were further complicated by the establishment of the water hyacinth, a Brazilian weed that made its way to the region by the 1910s, choking both small and large water bodies and paddy fields across the delta, particularly in those places where the problem of waterlogging was acute.

Between about 1900 and 1930, these two elements collectively brought havoc to agrarian production processes, contributing to reduced agricultural production output and increasing water-related diseases. As we shall see in the book, the impacts of these ecological changes were so severe as to call for reconsideration of the problem of poverty and starvation in the region from an ecological perspective. First, it follows that the great Bengal famine has to be seen from an ecological perspective; by considering the question of entitlement and access to ecological resources as well as to the market. Second, it is argued that the great Bengal famine has an ecological pre-history, implying that the problem of poverty and starvation started much earlier than the Great Depression of the 1930s. In other words, the great Bengal famine was as much socially constructed as it was the climax of environmentally-informed underproduction dating back as early as the turn of the twentieth century.

By the early twentieth century, the state was not only knowledgeable about the landscape of the Bengal Delta, it also had the political strength and expertise to appreciate the evils that accompanied the technology of railways and the biological invasion of the water hyacinth in the deltaic landscape of East Bengal. Yet why did the state fail to check their adverse impact? Did the failure emanate from the state's lack of control on the forces of capital or on its inability to effectively engage colonial science and technology or bureaucracy? Or was there a problem with the social and cultural perceptions of modern science and technology?[48] In our larger quest for narratives of relationship between ecological and social changes, we will try to address some of these questions in the book.

V

Samir Amin has recently remarked that the so-called less-developed countries are in large part 'destroyed by the intensity of their integration during an earlier phase of global expansion of capitalism'. He cites Bangladesh as an example of how this 'jewel of British colonization in India' has been destroyed though such integration.[49] This critique of the colonial and postcolonial phases of ethnocentric economic globalization in explaining the sorry state of many developing countries is important, yet in the specific case of colonial East Bengal, Amin's conclusion is only half true. This is because Amin, following colleagues in the world-system school, considers global integration as a linear process having the same impact in all places, and overlooking the fact that external influences only impact upon societies that are always changing through multiple endogenous social, economic and ecological engagements. This makes it difficult to measure historical development in a singular, progressive manner.

As we shall see in this book, a dynamic period in the agrarian economy, social formation, human well-being and political assertion by the peasantry took place in the nineteenth century. This was followed by a contrasting devastating decline by the turn of the twentieth century, culminating in the great Bengal famine of 1943 which killed about 3 million people. Yet the book is not about differences between the nineteenth and twentieth century, prosperity and decline, tradition and modernity, peasantry and the middle class, or Hindu and Muslim. It is a study of the way in which ecology helps to explain these categories, changes and differences.

Chapter 2 examines the ecological context of the limits of colonial control and the evolving patterns of agrarian relations in the nineteenth century. The next chapter examines the ecological context of the production of suitable crops, patterns of commerce and trade and connected issues of well-being and social formation. The Faraizi movement, described in the fourth chapter, developed not only as an Islamically-informed political resistance to colonial rule, but also as a sustained system of mediation and solidarity among peasants of diverse social and religious backgrounds, facilitated by local agro-ecological variables that they exploited effectively. Chapter 5 examines the motivations and patterns of gradual penetration into the primary production domain of a predominantly Muslim peasantry by the largely non-agricultural Hindu bhadralok. The sixth and seventh chapters focus on the emergence, growth and impact of the railways and the water hyacinth respectively, with particular attention to the way these were situated within the complex domain of colonial science, technology, bureaucracy and finance. Chapter 8 draws together the ecological and social changes since the early twentieth century to examine the famine of 1943 in its long-term perspective. The book concludes by summarizing the complex political ecology of colonial Bengal and by pointing to some of its implications for postcolonial Bangladesh.

2
Ecology and Agrarian Relations in the Nineteenth Century

The introduction of the Permanent Settlement in 1793 sparked intense debates in London and Calcutta on colonial governance and revenue management. For the next two centuries, from the work of contemporary figures such as James Mill and Ram Mohun Roy, these debates seem to have informed two strands of historiography of colonial Bengal. On the one hand, a large number of studies – looking at their subject from a variety of points of view: Marxist, nationalist and imperialist – trace the Permanent Settlement's mechanism, operation and impact on agrarian society and economy. On the other hand, a smaller number of works critically examine the ideological origin of the settlement within the context of the East India Company's governance ideas and practices.[1] Between the 'idea' of the Permanent Settlement and its perceived practical operations, there has been little focus on its limits in spatial terms. This lack of appreciation of the practical limits of the Permanent Settlement is less a result of the orthodoxy of certain historical approaches than the outcome of their inability to put ecology in its proper, long-term perspective. This chapter examines the ways in which the ecology of East Bengal thwarted, modified or muted the state's own standard mechanism for land revenue management, and the resultant impact on the access and entitlement of the peasantry to the region's agro-ecological resources.

'Wasteland' and the unmaking of the Permanent Settlement

During the intense debates on the modalities of the Permanent Settlement, its framers, including John Shore, attached importance to the wastelands in two different ways. First, the zamindars were incentivized

18

to cultivate wasteland that fell within their permanently settled estates. Second, it was agreed that those wastelands that were not included in the Permanent Settlement in 1793 be reserved as Crown land and 'as a source of income in the future'.[2] The general point is that the boundary of estates was so vague in 1793 that the East India Company could very quickly decide on reassessment and resettlement of the wastelands. The emphasis on keeping the wasteland under government control for securing future income for the state became all the more necessary when the much advertised prospect of a healthy revenue income through the Permanent Settlement gave way to what James Mill termed a 'permanent deficit' in the company's revenue. By 1797 rapid accumulation of the company's debt 'exceeded all former example'. The emerging critique of the settlement as a limiting force on the state's revenue earning was further informed by the 'larger loss' due to inflation.[3]

Criticism of the settlement on financial grounds was backed by a set of functional considerations. There were common expectations that the settlement would secure the financial interest of the company, restore order in its revenue management, create a benevolent landlord class and bring peace and prosperity to the peasantry. All these hopes, albeit mutually incompatible, were roped so tightly to the idea of the Permanent Settlement that any revelation of its faults and weaknesses was met with a great sense of alarm. By the turn of the nineteenth century, company policymakers were convinced that the Bengal Permanent Settlement was defective; these defects were found to be 'grave', 'serious' and 'enormous'. Policymakers thus began to look for an alternative revenue settlement by which the 'practical inconveniences' of the Permanent Settlement could be avoided, its 'mistakes', 'evils' and 'injustices' eliminated.[4]

In unofficial circles, the criticism of the settlement took a different form. One 'Civis' suggested in 1842 that the Permanent Settlement was too sophisticated to suit the local people who were not 'in an advanced stage of civilization and did not want a system more forward or refined than themselves'. He observed that 'their own homely institutions should have suited them better than the artificial version which we undertook to prepare for them. For every day purposes, an instrument may be too highly polished as well as too rough – too light as well as too heavy for the hand that is to use it. The blade of a razor is not suited to the handle of an axe.'[5] Civis, however, maintained that the colonial power considered 'abstract excellence instead of practical utility' and that the utmost they should have attempted was to improve the system of the country or 'even left it as it was'. Referring to the fact that the Permanent Settlement was losing ground – and losing supporters – Civis suggested that 'some retreat must eventually be attempted', particularly in the tracts not yet settled, thus preventing the government from violating 'the rights of their subjects by the establishment of the *Zemindary* system'.[6] About the same time, another unidentifiable reader of *The Times*, 'Libra', felt that the attempt to implant the English system of landed aristocracy through the Permanent Settlement had 'entirely failed' in that it was made hastily and without sufficient regard to the right 'as sacred and as ancient as those of the Zamindars'. He was satisfied that the government had 'postponed their consent to a permanent settlement in certain districts because they would not run the risk of giving perpetuity to injustice'.[7] The settlement was also criticized from a juridical point of view. Noting that it failed in any degree to improve the condition of those who cultivated the land, Raja Ram Mohun Roy suggested that the protection of the poor lay in 'disarming revenue officers of magisterial power, which was an essential part of Cornwallis's plan, and the immediate investigation of charges against revenue officers by the judicial courts'. While noting that some recent initiatives made by the government in India would address some of the existing problems of the revenue system, Roy nonetheless felt that the government did not go as far as to separate the magisterial from the revenue functions nor did it make essential changes to the system, which, he thought, would require 'a revision of the permanent settlement itself'.[8]

Criticism of the Permanent Settlement reached full circle by the 1830s, leading colonial policymakers to take different approaches to revenue settlement in different parts of India, including Madras and Uttar Pradesh. Interestingly early critics of the Permanent Settlement failed to see that it had been discarded in parts of eastern India as early as the year of its introduction. It was calculated by Cornwallis and Henry Colebrooke that no less than a third of the East India Company's territories, that is Bengal, Bihar and Orissa, was 'jungle'. According to Colebrooke, this estimate included lands fit for cultivation and excluded lands that were either barren or irreclaimable such as rivers or lakes.[9] The Sundarban forest tracts were kept outside the jurisdiction and mechanism of the Permanent Settlement and a separate office, called Commissioner of the Sundarbans, was created in 1816 for managing its revenue. The Office of the Collector, which played a key institutional role within a revenue district under the Permanent Settlement, had no other connection with the affairs of the Sundarbans than that the Sundarbans' revenues were paid into its treasury.[10] Later, following disputes between the government and the zamindars, who wanted a share in the forests adjoining their estates, the government clarified its position through further regulations.

Regulation III of 1828 proclaimed that the Sundarbans were the property of the state and had in no way been alienated to zamindars under the agreement of the Permanent Settlement.[11] Regulation XXIII of 1837 established the government's 'inherent title to share produce of all lands cultivated in the Sundarbans on the ground that the tracts were waste in 1793 and thereby not included in the permanent settlement'.[12]

The same policy was applied to other forest wastelands in East Bengal. For instance, the district of Chittagong was ceded to the East India Company in 1761 and there was a general survey of the district in 1764. When the Permanent Settlement came into operation, only the areas surveyed in 1764 were brought under its jurisdiction. Lands that were not included in the chitha of 1764 and lands reclaimed since the Permanent Settlement were considered the property of the government, which had the right to settle them with persons in possession.[13] The state was considered the 'principal zamindar' of the district – the area held by it being 'infinitely larger than that of permanently settled' area, while the revenue of the non-permanently settled portion was about 40 per cent of the whole.[14] Given the 1794 estimate of the total area of 9567 square miles in the districts of Chittagong and Comilla, of which 5250 square miles were uncultivated or waste,[15] it might be safely inferred that more than 50 per cent of the land in these two districts remained outside the jurisdiction of the Permanent Settlement.

In addition to the forest, another ecological feature of eastern Bengal soon came to be included in the colonial definition of the 'waste': chars and diaras. As the maps of James Rennell, drawn between 1764 and 1772, were being updated it soon became clear to the administration that huge amounts of new lands had been formed on the beds of the rivers and in the coastal regions. Like the forest wasteland, the chars and diaras were also kept deliberately out of the purview of the Permanent Settlement. Estimating the extent of the chars and diaras would anyway be difficult, since the rivers of the delta kept forming lands on an unpredictable scale and pattern. In the districts of Jessore, Faridpur and Barisal, new chars were forming constantly and islands were becoming part of the mainland, thus increasing the territorial boundary of the delta (see Figure 2.1). In the 1870s there were at least 133 such islands in Barisal.[16] It was estimated that in the district of Barisal, the natural aggradation increased the land area by 18 per cent or 1399 square kilometres between 1793 and 1905. On average, the settled area of the district expanded by 36 square kilometres per year.[17] In Noakhali, there was a 'multitude' of chars. In the three decades between 1793 and 1827 eighty-two chars were formed in the district of Comilla alone.[18]

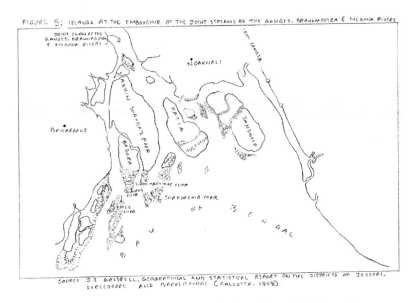

Figure 2.1　Drawing of coastal islands at the Bay of Bengal in the mid-nineteenth century

Source: G.E. Gastrell, *Geographical and Statistical Report of the Districts of Jessore, Fureedpore and Backergunge* (Calcutta, 1868). Redrawing courtesy: Rizwana Siddiqua.

The continuous process of land formation kept the local administration alive to the prospect of its utilization, which fitted the government's general policy of shifting away from the Permanent Settlement.[19] In 1811, the collector of the Comilla district, for instance, reminded the Board of Revenue of the existence of fine rice-producing chars that were being formed in the district 'daily', as being fit for immediate cultivation. He advised the board to take control of all unclaimed char lands with view to putting in farm leases immediately.[20] Under regulation XI of 1825, the government brought all such lands under its possession as khas mahals.[21] Henry Rickett, in his Settlement Report of 1849 on Chittagong, ascribed many of the problems of litigation to the 'ill-arranged and worse conducted inquisitions in order to "discover new lands"'.[22]

It was estimated by Baden-Powell in 1892 that 15.2 per cent of the land of greater Bengal remained outside the boundary of the permanently settled areas. This estimate does not seem to be reliable since it was based on the amount of revenue, the rate of which varied from estate to estate and even from plot to plot. The original source of this estimate was the annual reports of the *Revenue Administration of the Lower Provinces* (later on, Bengal), which record revenue returns with specific reference to 'permanently settled' and 'non-permanently settled' areas. But the reports do not specify the extent of areas under the different systems of revenue administration, though the number of estates is provided.[23] Even if we accept that 15 per cent of all revenue-paying land in greater Bengal lay outside the jurisdiction of the Permanent Settlement, the percentage would be far larger in East Bengal which possessed comparatively more wastelands in the forms of cultivable forests and existing and continually emerging chars and diaras. At the same time there were lands which were 'permanently settled' – but not in Cornwallis's sense – in that the settlers were non-zamindars, ranging from talukdars to raiyats, who were attached to 'permanently settled land', but under the authority of the government.[24] By the turn of the century, it had become clearer that amounts of rent from khas mahal areas were proportionately larger than those from permanently settled areas. In 1900, a fourth of the land revenue of Bengal came from non-permanently settled areas.[25] Since such khas mahals predominated in East Bengal, the ratio here would have been much more strongly in favour of the non-permanently settled areas. Considering the extent of the non-permanently settled areas and the continuous formation of new land, as well as the 'pseudo-permanent settlement' areas, we may assume that at least 50 per cent of the territories of eastern Bengal remained beyond the direct jurisdiction of the Permanent Settlement.

As the ecology of East Bengal encouraged the colonial state to deviate from the standard mechanisms of revenue management, so it also weakened the economic prospects of the zamindars. Erosion by the deltaic rivers reduced the extent of property held by the zamindars. Rivers also created land in the forms of chars and diaras offering the zamindars the opportunity to recoup their loss by claiming ownership of these new lands. Yet, under Regulation XI of 1825 proprietorial interests in areas of an estate that had been eroded by rivers were 'lost forever', and such lost areas would not come back under their control even 'should land reform on the exact site which their estates occupied'.[26] Although these regulations entitled the zamindar to an exemption from paying rent for the diluvated property, some zamindars continued to pay revenue for the eroded estates in the hope of obtaining pre-emptive ownership of the land which might reform on the site they once occupied. But the administration asserted that this was 'an entire misapprehension in the existing state of the law'.[27] The interests of the zamindars were further jeopardized by nature's crude indifference to their cause, as exemplified by the fact that in Noakhali, the accretions to permanently settled estates between 1863 and 1881 were 'very small', while the accretions to non-permanently settled estates were 'considerable'.[28]

In most cases, the zamindar would stop paying revenue for a property which had disappeared under water. As soon as the zamindar ceased to pay the revenue, the estate was put up for sale in order to realize the arrears. Since no one would bid for estates which no longer existed, the proprietary right was bought by the government for a nominal sum.[29] Besides playing destructive games with the landed property of the zamindars, the deltaic rivers also raised the beds of the beels that were situated either inside the permanently settled estates or beyond by depositing silt and rendering them cultivable. The absence of definite rules regarding the proprietorship of these cultivable wastes, particularly those adjacent to the permanently settled estates, created legal complexities that embroiled peasants, government and the zamindars. The government's efforts to bring under its control as much land as possible and the resistance to such efforts from the zamindars led the government to assert its legal rights over such lands and to recognize the rights of those raiyats who had settled and cultivated the same land. The initial legislative attempts to retain most of the wasteland both outside and inside the boundary of the Permanent Settlement culminated in the Tenancy Act X of 1859.

If Act X of 1859 bestowed upon the cultivators proprietary rights on land, it also cornered the policy of the Permanent Settlement and its beneficiaries in a number of ways. 'A Lover of Justice' (pseudonym of John Cochrane, an English zamindar) considered this act a severe blow to the Permanent Settlement, particularly in those areas where the process of land formation was in operation and where wasteland existed in plenty. He observed that by the drainage of villages, the decomposition of leaves and straw and the addition of manures, the low lands in this country were gradually rising, and he lamented that the raiyats were continually extending their boundaries by encroaching upon these waste lands. Cochrane was disappointed that rather than 20 years' uniform rent receipts, only receipts of a few years were sufficient in proving the proprietary rights of the raiyat in the High Court. In the opinion of the judges it was not even necessary for the raiyat to prove the quantity of the land he originally obtained from the zamindar by any documentary or other trustworthy evidence. He was simply to say that he had held so much land at such a jumma and his words were 'taken as gospel'. Cochrane also observed that if a hereditary zamindar had filed old chithas to show that the raiyat had added adjacent waste-lands to his existing land, his objections were noted down as 'frivolous and vexatious'. At the same time, since the High Court was generally in favour of the raiyats, the judges in the mofossal rarely offered any contrary verdicts. If they did otherwise, Cochrane observed, their decisions were sure to be upset by the High Court, and at the end of the year their promotions would be affected if a large number of their decisions had been overruled. In a show of utter exasperation, Cochrane claimed that if one of these judges had changed place with a zamindar for a year or two, or even acted as a superintendent in a large zamindari estate, he was sure the judge 'will recant his former errors and be in a position to give opinion worthy of all respect'.[30]

'Lover of Justice', with his eloquent criticism of the government's unsympathetic attitude towards the idea and institution of the Permanent Settlement, failed to convince the administration. As early as 1861, the government of India made it clear that in this region, where wastelands abounded, it found a 'clear gain to the revenue of the State' in the amount of land that came out of the transfer of the zamindari tenures to ordinary individuals.[31] For example, when Khajah Abdul Ghani, Nawab of Dhaka and one of the most influential zamindars of East Bengal, claimed a large territory on the bank of Phuljhur River in Barisal, his claim was rejected by the Privy Council and the area came under khas management.[32] In 1871, the practice of settling lands permanently within the non-permanently settled areas was forbidden and khas management was encouraged wherever possible. At this time the government refused to comply with the Landholders and Commercial Association of India's demand for repealing the existing laws regarding alluvion and diluvion. The government asserted that 'it was the bounden duty of the Government not to surrender one of the rights to alluvion which it holds as trustee for the public' and that 'no decision of the Privy Council can affect the validity of the law'.[33] Ascoli noted that the general principle of the government to encourage khas management was 'revolutionary'.[34] With the policy crystallizing in favour of bringing as much land as possible under government control, the territorial breadth and fiscal significance of the Permanent Settlement fell under increasing suspicion.

However we interpret the colonial administration's distancing of itself from the Permanent Settlement in the course of the nineteenth century, it was obvious that the central motif of the state was to appropriate agricultural resources more directly by reducing the hierarchy of mediation between the actual producer and the state. This was equally true for those utilitarian protagonists in the company's central administration and the local British officials who, in discouraging permanent settlement and favouring raiyatwari settlement, wanted to 'revitalize the rural economy by setting cultivating peasant brotherhood free from the depredations of corrupt state functionaries and greedy landlords'.[35] But the consequences of these efforts were not uniform across India. Although the administration's attempt to recoup the loss from a fixed revenue regime shifted its attention to the wastelands, it was difficult to find a coherent alternative to the Permanent Settlement precisely because of the fluidity and relative illegibility of the wasteland of the region. There were attempts to acquire knowledge of the subtleties of the landscape and to produce statistics by mapping and surveying as early as the mid-eighteenth century,[36] but it is doubtful if these 'scientific' endeavours made the state any more knowledgeable about the intricacies of the interior of the country. As Matthew Edney puts it, the horizontal spatial boundaries of the imperial space of India, which enclosed, divided and so gave political meaning to an otherwise homogeneous space 'merged imperceptibly with the vertical boundaries of the empire's social hierarchies'.[37] If this is true of the more settled regions which had been exposed to the Mughal imperial gaze for centuries, it would be more so for the sparsely inhabited wastelands of East Bengal. It would be interesting to examine how this 'imperceptibility', which arose at the interface between the colonial state and the fluid and forested frontier zone, informed the state's attempt to reproduce revenue and agrarian relations in the nineteenth century. One way of doing it could be to trace the reclamation process and resulting tenure arrangements in this region.

Patterns and agencies of reclamation of wastelands

The first reclamation process in the Sundarbans was initiated in the district of Jessore by its magistrate, Tilman Henckell, a few years after the devastating famine of Bengal in 1770. Henckell cleared some areas at his own expense and created three large markets (Kochua, Chandkhali and Henckellganj) on the banks of the rivers that connected Calcutta to East Bengal. In his reclamation scheme, submitted to the Board of Revenue in 1784, Henckell proposed to grant Sundarban plots to interested private agencies on favourable terms, including the option for the leaseholder to retain for himself as much as 200 bighas of land. There was provision for waiving the payment of revenue for the first three years; then it was supposed to be two, four and six annas to the rupees for the fourth, fifth and sixth year respectively. Finally, eight annas were to be fixed for the seventh and succeeding years. In support of this arrangement, Henckell argued that the scheme would not only prepare the wasteland for generating revenue but the reclamation and cultivation process would also form a 'grand reservoir for rice against seasons of drought or famine, as the crops grown in the Sundarbans were little dependent upon rainfall'.[38] The Board of Revenue approved the plan.

In the following few years, stray reclamation on Henckell's pattern took place with mixed success. In 1819 revised terms of lease were formulated in which no revenue was demanded for the first seven years, with payment thereafter gradually increasing to eight annas per bigha in the eleventh year, with the stipulation that the rate would be fixed forever from that year. But the conditions were that one-eighth of the leased land had to be cleared in three years. Not a single lease was made under these rules as the rent-free period was considered too short and clearing conditions too stringent. Another scheme followed in 1830. This time, a fourth of the leased area was to be excluded from assessment for ever. Revenue for the remaining lots was nil for the first twenty years. One hundred and ninety-five lots were leased out under the terms of these rules. A revised policy was announced during the viceroyalty of Dalhousie, who, in the context of financial constraints put on the government on account of a number of expansionist expeditions, paid special attention to reclamation. This 1853 scheme was essentially similar to that of 1830, but the rates of assessment were considered 'exceedingly liberal'.[39] The overall pattern of granting leases in the Sundarbans as formulated by Dalhousie, with only minor modifications, remained in place until 1903 when reclamation through large-scale grants was discontinued.

While the liberal terms appeared lucrative to the leaseholders, there were two aspects of the reclamation process that directly or indirectly left them in a disadvantageous position. First, while providing the lease on favourable terms, the government also ruled that failure to reclaim certain tracts of lands within a stipulated time would result in the termination of the lease, including those parts that had already been cleared and cultivated.[40] Under the rules of lease of 1830, a fourth of the total leased area was to be cleared and cultivated within five years – otherwise, the lease was to be forfeited and contracted anew. Although the rules of 1853 were similar to those of 1830 in many respects, a larger portion of the land leased was required to be cleared within a specific period.[41] Another associated problem from the point of view of the leaseholder was that the government sold the 'fee simple' options to the leaseholders in Arakan, Assam, Cachar and Darjeeling, but not in the Sundarbans. It was argued that, as the main reason for the change of terms of the grants in 1853 was to ensure the clearing of the jungle, the Board of

Revenue could not deprive the government of the power of resumption on failure to clear the forest because this 'could not well be enforced if the fee simple of the land were parted with'.[42] These constraints put the Sundarban leaseholders, unlike grantees or colonizers in the Nilgiri Hills or Darjeeling, under considerable pressure and compelled them to give leases to the abadkar or actual reclaimer-peasant on the same lucrative terms that they themselves were given by the government. The leaseholder's conceding of generous terms to the abadkar was further informed by the way jungle grew up in the fertile deltaic landscape. Hunter noted that it was difficult for a grantee or leaseholder to replace a tenant who had left him and that owing to the rapidity with which jungle sprang up in the Sundarbans, a grantee could not 'afford to allow his lands to lie fallow'.[43] There was also the question of wild animals which had to be tackled by the abadkars.[44] By the turn of the nineteenth century, such was the state of wilderness that to move with 'any tolerable degree of safety from Calcutta to any of the adjacent districts, a traveller was obliged to have at each stage, four drums and as many torches'.[45]

Chars and diaras were leased either to the neighbouring zamindars, both local and European, or more often let directly to the raiyat on very lenient terms. If a contract was made between the government and a zamindar, the latter would give a hawala or commission to a man of 'energy' or 'capital' on favourable terms to bring the land into cultivation. The hawaladar cultivated the land with the help of neighbouring raiyats, and eventually developed it as abad. The hawaladar asserted his leadership as he improved new lands under the original temporary contract and eventually found himself in possession of full talukdari rights. If a contract was made directly with the raiyats, there was no intermediary between them and the government. They were granted land at a rasadi or progressive rent for the purpose of bringing wastelands into cultivation. A nominal rent or no rent was payable for the first year and the rent increased progressively with the development of cultivation. The rights of under-tenants in the alluvial lands were defined under Regulation XI of 1825, and in the subsequent regulations of 1847 and 1858 these rights were reasserted. Webster recorded that chars were given to the cultivators on generous terms and that government policy was increasingly directed at contracting with raiyats instead of with hawaladars, which had been the case in earlier days.[46]

There were two ecological peculiarities of chars and diaras which also favoured ordinary cultivators. First, as noted above, while the diluvial actions of the rivers, which cut away tracts of land, caused great loss to the zamindars, they were not compensated by being granted lands formed through alluvial accretions. Instead these lands generally came under direct government control and were assessed anew.[47] Many zamindars were ruined by such random acts of nature whereas the newly created lands, being now government property, came into the possession of the abadkar. Second, these chars and diaras were frequently neither measured nor judicially settled because of the 'extraordinary difficulty' in measuring the littoral districts. Thus in the 1880s, the Board of Revenue asked for a report whether in the Pabna bank of the Ganga there was 'any good reason for making a re-survey so soon again, the last survey having been made only eleven years ago'.[48] The same loose application of measurement followed in other districts. While by 1880, Dhaka, Pabna, Barisal, Faridpur and Comilla districts had been surveyed only once, Mymensingh was not surveyed at all.[49] As for the Chittagong district, there was 'no single khas mehal' which had 'ever been completely surveyed'.[50] During the long gaps in systematic survey and settlement, tenants gathered to settle and cultivate. The zamindars would also try to avail themselves of this opportunity of claiming authority over these lands, heightening tension between themselves and the raiyats. The raiyats often won the day, as the zamindars were not in a legal position to exert their authority on such lands.

As a result of the irregular and infrequent measurement of lands, land administration was invariably conducted on the basis of incomplete knowledge. This, along with associated legal complications, was often the cause of insecurity of investment of large leaseholders. For instance, Baboo Kally Mohan, a leaseholder who was also a High Court pleader and in this capacity was 'more competent to fight in law courts than the ordinary class of men who took such farming leases', had asked the government repeatedly to have his lease cancelled. A superintendent of the Dearah survey noted that the circumstances were such that it was almost hopeless to effect a farming settlement of those lands where the proprietors remained reluctant. It was therefore necessary, he suggested, that after the expiration of a reasonable period of time, land on which the proprietors did not pay rent might be settled with another leaseholder or held under direct government control.[51]

Generally the leaseholder was thoroughly dependent on actual reclaimers. The hawaladar held lands and paid rent for a village, or half a village, but he could not clear and cultivate the entire portion that he had received from the leaseholder. He would therefore create sub-tenures. For instance, a hawaladar created within his own tenure a subordinate tenure called nim hawala and subsequently an ausat hawala, intermediate between himself and the nim hawaladar.[52] Neither the government nor the leaseholder interfered in the hawaladar's right to create sub-tenures as all parties involved recognized that without such a process the project of reclamation and cultivation would not come to fruition within the scheduled time. This system of sub-tenures reflects a considerable degree of dependence between the tenure-holders and sub-tenure holders in the face of the enormous risk, hardship and difficulties involved in the reclamation process. Thus though the leaseholder was apparently offered a generous reward for prospective reclamation, he had to operate in a difficult area between meeting the demands of the government and getting the clearance done on time in difficult terrain. All these factors informed the pattern of the leaseholder's engagement with the abadkar, without whose support at the field level reclamation seemed impossible. The result was that not only were settlers able to obtain very favourable terms of rent, they were often assisted by the leaseholder with money advances, or by the purchase of cattle.[53] When the land was sublet in bhag or shares, the landholder received half of the produce from his sub-tenant in lieu of rent. The oxen for the plough were supplied by the landholding raiyat, but seed grain, plough, and all other necessary agricultural implements were furnished by the sub-lessee.[54]

The flexible low-revenue regime and logistical support allowed an emerging band of cultivators to make significant gains. Yet given the level of risk and physical hardship which the reclaimer-cultivators had to endure, they wanted a clear register of rights on the land they reclaimed. As the cropping of revenue off the wasteland was the key aim of the colonial state, it was not a problem for the administration or the leaseholders to accommodate such demands by the abadkars. Such concessions contributed to the emergence of a vast majority of cultivators who came to be known as 'occupancy raiyats', or middle raiyats, who played a dominant role in agrarian relations in the nineteenth century. In the next section, we will examine the evolution of this category of peasantry in some detail.

The reclamation process and rise of the occupancy raiyats

The claim to occupancy rights was based on abadkari swatwa, or right of cultivation, which was founded upon 'original reclamation'. According to Westland:

> The first and heaviest part of the clearing of any plot of land is usually done at the expense of the proprietor, the person who has settled

with Government for the land; and when the clearing has proceeded to a certain point, he settles ryots upon the lands thus partially cleared and they bring it into cultivation. These ryots call themselves 'abad-kari' or reclaiming ryots, and esteem themselves to have a sort of right of occupancy in their lands.[55]

Robert Morrell, holder of one of the largest grants in the Sundarbans, used to contract hawaladars at a certain rate. During the settlement process the abadkars were given amalnama, or orders to remain in possession and to cultivate. Under this engagement the hawaladars were obliged to clear the forests for cultivation and settlement. On the question of whether the abadkars would take up pottahs (written rights of possession) unless given in perpetuity, Morrell mentioned that some would come and work without a permanent right of possession, 'but', he continued, unless he gave such pottahs, 'they would not come willingly and in numbers' and that his object was to get as many raiyats as he could.[56] It is noteworthy that in the case of cancellation of a lease or changes in the ownership of an estate, for whatever reasons, the abadkhars' proprietary rights were not affected.[57] Under Article V of the term of lease of 1853, when a grant was resumed, the land found to be actually under cultivation was to be measured and settled with the cultivators or under-tenants.[58] The government policy of safeguarding the occupancy of raiyats continued in the succeeding enactments and policies until the late nineteenth century. For instance, following the requisition of land for the Matla Port project in the Sundarbans, the Board of Governors expressed their hope that justice had been shown to the raiyats and others who had 'suffered to acquire rights of occupancy' over relevant possessions, subsequently as well as previously to the purchase of the lot by the government.[59] This project, however, failed and the government resumed a large number of lots that had been under the control of the company concerned and these lands were settled directly with abadkars.[60] Dampier, a commissioner of the Sundarbans, noted that reclamation of land was the source of proprietary rights in eastern Bengal. Referring to the reclamation process in both Chittagong and the Sundarbans forests, Dampier suggested that the reclaiming abadkars, whether authorized or 'trespassers' and 'usurpers', were considered 'as friends and not enemies to progress and public wealth'. He asserted that as he had always done in the Sundarbans, so in Chittagong he was an advocate for a 'liberal policy' as to recognizing or conferring upon the abadkar rights on their land. He suggested that since 1848, the law had given definite rights to the occupancy raiyat and suggested that superior landholders in Chittagong 'must accept this'.[61]

The Bengal government's singular emphasis on the extension of cultivation and the giving of institutional, legal and moral support to the abadkars at the expense of the zamindars was not left unchallenged. 'Lover of Justice' thought that under the operation of Act XIL of 1841 and Act I of 1845, the simple khodkasht (resident) raiyats settled in an estate since the Permanent Settlement had no rights of occupancy against sale purchasers, and that paikasht (non-resident) raiyats were 'nowhere expressly mentioned' in the laws referring to Bengal. Therefore, he argued, the raiyats never dreamt of obtaining the right of occupancy by simple possession for 12 years and that this right had been 'thrust upon them unasked and unsolicited'. He further noted with surprise that the legislature not only gave this right to 'every new squatter residing in a village, but extended the privilege to all Pycust Raets [Paikasht raiyats]'.[62] However, in a series of correspondence, the government argued that 'there were no reasons why in the interest of the landlord Government should set itself to defeat privileges which the policy of the law and the custom of the country give to raiyats; the law and custom being probably founded on the sound principles and being for the interest of both parties'. The government also felt that in a country where the landlord neither found the capital nor made the improvements for ordinary agricultural operations, the right of occupancy enabled the tenants to do so and that the accrual of such rights could not really 'materially detract from the value of the estates'. At the end of the exchanges, it was asserted that 'there will now be no misconception on the part of any of the officers of this Government as to the order that they must not interfere to defeat occupancy rights without special reasons'.[63]

In the context of the debates as to whether the process of accrual of occupancy raiyats should or should not be left to continue, the government of India advised the government of Bengal to the effect that until the former had considered the whole question of occupancy rights, no lease should be granted 'without provision which shall prevent occupancy rights from growing up in virtue of any occupation under such lease'.[64] Disregarding this advice, the government of Bengal replied that the terms of Act X of 1859 were most express and explicit in that 'every' raiyat had a right of occupancy who had held land for twelve years 'whether it be held under pottah or not'. It added that the decision was not greatly opposed by the zamindars who only cared for an increase of rent from time to time and that they had never 'shaped their pottahs with a view to reserve a right of re-entry'. The decision of the High Court as well had made not the 'least difference' in this respect.

Arguing in favour of occupancy landholding, George Campbell, Lieutenant-Governor of Bengal (1871–74), found two different relations between landlords and tenants. By far the oldest and most usual was that while the landlord (either zamindar or the state) was entitled to the rent, the tenant cleared the waste, prepared the land, built the houses, provided water and drainage, and did all that was required to make productive the soil which 'God had given in an unproductive state.' Under such circumstances it was always the law or the custom or the understanding that the cultivator was not to be disturbed so long as he paid a fair rent, or that if he was disturbed he was to be fully compensated for his interests. A variation on this was that the landlord prepared the soil and, having made the land fit for cultivation, let it to a tenant only on contract basis, the latter being subject to eviction after the expiration of this term. Campbell believed that 'there had not existed in the world a country where the tenant prepared and stocked the farm and was yet liable to be evicted at pleasure'. According to him, India had not yet reached or in any degree approached the stage when the landlord prepared the farm. It was also noted that rightly or wrongly the British government in India had everywhere conceded rights of some sort to the parties with whom it dealt for the land revenue – to the raiyats of Madras and Bombay (even to those raiyats without any previous rights), to the village holders of northern India, or to the zamindars of Bengal. Why then should they 'jealously guard against the concession of the lowest scale of privilege to the ryots of the khas mahals of Bengal?', Campbell asked. In fact, the government of Bengal thought it necessary to prevent any attempt on the part of the government [of India] 'which might construe as setting an example of systematic attempt to defeat occupancy rights'. It was, therefore, suggested: 'it would not be desired to entrap the ryots who already possess rights of occupancy into signing documents which would destroy their subsisting rights'.[65]

George Campbell's attitude towards occupancy rights seems to have been informed by the agrarian relations that emerged independently of the influence of the Permanent Settlement. For, as he recollected later in his autobiography, in Chittagong and in Sylhet, for instance, things were working on what was almost a raiyatwari system, land being held by small peasant-proprietors, with the 'permanent fixing of land revenue' a measure that was 'past and gone'.[66] Revealing as this was, even in areas under the Permanent Settlement, where considerable wasteland existed the zamindar had difficult relations with the reclaiming raiyats. In most cases, when a zamindar patronized reclamation of wasteland in his own estates, he appeared to be more liberal to the reclaiming tenants than to the tenants in the more settled tracts. 'Lover of Justice' believed that lands were let to raiyats on the 'express understanding' that they were to hold the same from year to year 'during the pleasure of the landlord' and that whenever the market nirick rose, they were either to pay the enhanced rate or quit the land – very seldom, if ever, were pottahs or kabuliat exchanged on such occasions. But he also acknowledged that pottahs or kabuliat were exchanged when jungles were to be cleared, or wasteland of long standing was to be cultivated at a reduced permanent rate.[67]

Official correspondence between the zamindars of eastern Bengal, headed by Khawja Ahsanullah, and the government of Richard Temple (1874–77), who succeeded George Campbell, throws some light on the legal rights of both ordinary occupancy peasants and the zamindars. In a letter to Temple, the zamindars pointed to their relative disadvantages vis-à-vis the occupancy raiyats on a number of issues and demanded remedies to that effect. First, it was argued that the gradual increase of the value of produce, and the consequent reduction of the value of money, had made the zamindar poorer. Second, the zamindars lamented that the attempt to enhance the rate of rent (to prevent them from becoming poor) was largely foiled by the difficulty of proving enhanced rates of revenue where the raiyats were 'refractory' and by the 'bias of most of the Courts against enhancement suits'. It was noted that in the case of a zamindar refusing to accept a tenant's tender of arrears, the law provided that a tenant could deposit the money without any action being brought in court, and thus 'save himself from all liability whatsoever'. Besides, the zamindars explained, suits for arrears were very expensive, and it was almost impossible to realize the rents by suits because the raiyats united against the zamindar, with the consequence that a very large number of suits had to be instituted. On the whole, the zamindars could manage only 'a very disproportionately small percentage' of law suits in their favour. But, even where a decree was obtained for arrears due, the zamindar very often did not obtain the whole of the actual costs of litigation; and the low rates of interest allowed on decrees were 'by no means calculated to induce the ryots to satisfy the decrees at once'. In the context of 'enormous difficulties' in realizing arrears of rents 'at even the existing rates', the zamindars left the question of enhancement aside and clamoured for 'a stringent law for the absolute forfeiture of rights of occupancy on non-payment of arrears of rent'. The zamindars also pointed out that though the principle of limited relief against forfeiture for non-payment of rent introduced into Bengal by Act X of 1859 was recognized in England, it would not be equitable to give this relief to recusant raiyats in Bengal where, unlike in England, zamindaris were often sold due to the 'sunset law' and for the default of the raiyats. The zamindars also argued that the 'fear of absolutely losing their rights of occupancy will induce the ryots to pay in their rents in time'.[68]

In response to this petition, the Bengal government informed the zamindars of its inability to accede to their request of scrapping the legitimate rights of the occupancy raiyat. Instead, Temple congratulated the zamindars on the increase of occupancy raiyats as he felt that the increase of tenants' rights was 'good for them [zamindars]', as it gave the cultivators 'an interest in the land, and in effecting improvements which are ultimately beneficial to the landowners among others'.[69] The question of the actual cultivator's occupancy rights on land was further consolidated by the Tenancy Amendment Act of 1885.

Obviously, not everyone who cultivated land in East Bengal was an occupancy raiyat. This was partly because participation in the reclamation process was taking place on a 'first come first served' basis, and in particular towards the later nineteenth century, when lands were becoming scarce and the population growing faster than could be accommodated on new and existing lands, later arrivals might have been less privileged. In fact, there seemed to have developed different categories of raiyats: rent-free or permanently-fixed holders, occupancy raiyats, and non-occupancy under-raiyats or tenants-at-will. But, in the context of the nineteenth-century ecological conditions of the East Bengal delta, the first and the fourth category of raiyats were a minority.[70] At the same time, there appeared little qualitative difference between occupancy raiyats and the rest, as far as de facto rights over land were concerned. Even if we consider that non-occupancy raiyats existed to a certain extent, economic and social differentiation between this category and the occupancy raiyats was probably less in the context of the wilderness, which exposed all parties involved in the cultivation or reclamation process to considerable risk, than might have been the case elsewhere. For instance, by the 1870s the holders 'were protected from ejectment, even in those cases where such leases were granted expressly for a term of year only'.[71] It was reported that cultivators would seize upon fragments of land here and there, which were easy and profitable to cultivate, and would then refuse a lease unless they were offered favourable terms of contract.[72] In the coastal regions, the under-raiyat held from generation to generation under regular pottahs granted by his raiyat-landlord, and he acquired right of occupancy with respect to his holdings. He could not cut ponds, but could do all other things on the land he occupied, including cutting trees.[73] In December 1871, the full bench of the High Court found unanimously that the specification of a term of a year did not take away the accrual of the right of occupancy.[74] Meanwhile the government's suggestion was that 'in ordinary cases of ordinary agricultural ryots', special means were not to be taken, nor special clauses to be inserted in order to prevent the accrual of occupancy right.[75] It appears that tenants-at-will too had the opportunity to join the category of occupancy raiyat, though this option was gradually becoming more limited by the turn of the twentieth century, as the ecological regimes of new lands and forests were becoming scarcer.

The question of rent

Occupancy raiyats generally enjoyed a fixed rent. The under-raiyats or tenants-at-will were theoretically subject to flexible rents and eviction in case of failure to pay. But as long as new lands were available for reclamation and cultivation, the question of rent collection often appeared to be informed more by the difficulties in the reclamation process in the deltaic wilderness than by a differential rate of rent to be collected from a hierarchy of the peasantry. By Regulation V of 1812, parties were given liberty to make their contract on any terms, provided that no contract for irregular taxes was to be considered legal.[76] In Kutubdia, an island between coastal Chittagong and Noakhali, the rent rate had not changed since 1835, and a large remission had also been granted. As early as 1826, the collector of Chittagong, while justifying his lenient assessment to the talukdars of various mouzas of Taraf Johana Fernandez of this island, expressed his 'firm belief' that if a revenue agent was allowed to exact as much as he could from the under-tenants, a large part of the estate would 'speedily be deserted'.[77] It was reported that the new settlers in the island of Hatia paid no rent for eight years, thereafter the rent increased progressively with the value of cultivable land.[78] Following the expiry of a farming lease in Chittagong in 1875 and a devastating cyclone in 1876, many of the villages were depopulated. As the revenue decreased, the government again started offering inducements to the raiyats. No leases were given, but the raiyats cultivated as trespassers without paying rent.[79] In no estate in Cox's Bazaar, for instance, was there any systematic plan for revenue management and collection. A more common practice was to keep account only of the demand of collections and balance of rents, and leave 'tenants and outsiders to alter boundaries and to encroach or to squat on waste lands without any permission and without payment of rent'.[80]

From the government's point of view, a particular problem linked to the fluid environment of the delta was the difficulty in the measurement of land and its settlement. Ulrick Browne, the commissioner of Rajshahi, advised the government against carrying out a proposed annual survey and settlement of lands, arguing that survey initiatives might give a false impression of enhancement of rent and thus prevent the 'ignorant' raiyats from bringing lands under cultivation, while as things stood they knew that measurements only took place at long intervals, and that they could 'enjoy the fruits of their industry for some years without paying increase of rent'. In response, the Director of the Agriculture Department criticized Ulrick for contemplating a system of management under which raiyats might 'encroach upon and cultivate as much fallow as they please, in the intervals between one decennial of fifteen-year settlement and another, without payment of rent for it'. However, he endorsed Ulrick's ideas as far as eastern Bengal was concerned and noted that whereas it was a principle which a Bihar zamindar would regard with astonishment and alarm, 'from the abundance of fallow land' which was 'awaiting the plough' in the deltaic regions (in this case, parts of Rajshahi division), it might 'perhaps be recognized [there] without much harm'.[81] Even in those cases where the government was aware that revenue paid by the raiyats could be increased, efforts to get an enhanced rate were not always strenuously made. This relative leniency was sometimes informed by the fact that the raiyats were 'most tenacious' of the supposed inviolability of their existing dowls, which they considered 'fixed in perpetuity'. In such cases the government usually took exception in pressing its legal rights as it considered it 'morally and politically inexpedient'.[82] It appears that some of the under-raiyats, who officially lacked occupancy rights, actually shared the privileges of an occupancy raiyat, at least as far as enjoying a fixed or nominal rate of rent was concerned.

Between the tendencies of the raiyats to avoid paying rent and the administration's indifference to the clamour for rent enhancement by the landlords or large leaseholders, there arises one important question: what were the patterns of rent relations among the cultivators themselves, particularly in the cases where tenural hierarchy differentiated the cultivators? In some parts of northern Bengal, jotedars or substantially rich peasants were present in the early colonial period. With plenty of land and with occupancy rights, they did not even bother about physical participation in reclamation, but the picture seemed to be different in the active deltaic areas of East Bengal. Leadership in the reclamation process was of course crucial, and talukdars or hawaladars assumed the leadership and enjoyed considerable authority over ordinary raiyats. But given the environmental circumstances in the delta, it was probably not the capacity to collect rent from those lower down the hierarchy or corresponding sub-tenants but the capacity to employ physical enterprise and labour which held intra-tenural relations in this system intact and which rewarded each party involved. At least until the early twentieth century no remarkable differentiation was evident within the society of peasant small-holders.[83] Ecological constraints bound all incumbent parties by a common thread of agrarian relations. All the parties involved – the government, the leaseholder, the peasant leader of the reclamation process and the ordinary peasant – wanted to reclaim new lands from chars and forests, but each of them was constrained by the practical difficulties involved in the process. The government employed its authority, the leaseholders their capital and the peasants their labour. In the end, the peasants proved to be successful not only in getting proprietary rights to the land they reclaimed, but also in achieving restraint when the state and leaseholders pursued rents. It was not surprising that in eastern Bengal, on average, rents remained low in comparison to other regions in eastern India.[84]

This chapter has shown that although the East India Company introduced the Permanent Settlement in most parts of eastern India to ensure a fixed, constant flow of revenue, it did not consistently impose the system on the fluid and forested frontier of East Bengal. This was because the colonial state perceived the reclamation of the forests and other wastelands as an opportunity to expand its dwindling revenue base, the result of a permanently fixed ceiling on revenue income. The state was compelled to attract settlers to the new frontier of cultivation by offering low rents and favourable terms for the tenancies of government estates. The tenancy legislation of 1859 and 1885 consolidated the actual cultivator's occupancy rights. This narrative illustrates the ways in which the ambitions of the colonial power, either in its everyday structural practices or through its ideologically-informed policies, were greatly compromised in relation to the wilderness which it categorized as 'wasteland'. However, the aim of this chapter was not to contribute to the debates around the question of the extent of colonial power and influence in India that have taken place between what are known as the Cambridge and Chicago schools of historians. Rather this chapter has set a background against which we can examine how the specific agro-ecological space in East Bengal brought the state and the peasants into closer negotiating positions, facilitating larger social and political developments. The next two chapters focus on these themes.

3
Economy and Society: the Myth and Reality of 'Sonar Bangla'

The colonial state's liberal attitude towards the peasants in East Bengal, as described in the previous chapter, was largely informed by the inaccessibility of the region's fluid deltaic frontier. But ecology was also significant in the way it induced the state to maximize and commercialize cultivation for supplying the domestic and international market. It is important to examine the way in which agrarian society responded to the complex situation emerging through the state's inability to restrict the peasants' access to ecological resources on the one hand, and its compulsion to generate tax and trade on the other. This chapter examines these issues through focusing on three broad themes. First, it looks at the way in which the ecology of the region provided for the production and commercialization of crops that were acceptable to both the peasants and the state. The second set of issues relates to the impact of ecology and the commercialization of agriculture on the economic and physical well-being of the peasantry. Third, the chapter will examine the process by which agricultural development and human well-being influenced agrarian social formations in the region. This approach should help unravel the myth and reality of 'Sonar Bangla' in the colonial era.[1]

Ecology and crops

One key ecological factor underlying the productive capacity of East Bengal throughout the nineteenth century was the fluvial strength of its rivers, which offered two important eco-system services.[2] First, silt deposited by the rivers, in addition to creating new land, continued to fertilize vast catchment areas. Second, the sheer volume and velocity of the rivers that drained to the Bay of Bengal kept the salty seawater at bay, sustaining a fresh-water ecosystem.[3] The influence of the river system in the creation and fertilization of land was of considerable importance in territories stretching from Rajshahi in the north-west to eastern Jessore in the south and throughout all of eastern Bengal, with the exception of two laterite patches of high land around Dhaka and Rajshahi. In the mid-nineteenth century Joseph Hooker found it remarkable that in the coastal district of Noakhali, the soil yielded plenty of salt while the water of the Meghna that flowed through it was not in itself saline, but was only salty as the result of repeated inundations and surface evaporation. Hooker also observed that fresh water was found 'at a very few feet depth everywhere', though it was not very good.[4] In addition to the fluvially vibrant rivers, an optimum monsoon rainfall sustained the freshwater ecosystem in the region. Whereas the annual average rainfall was between 50 and 60 inches in the district of Bankura of West Bengal and about 60–75 inches in the central district of Nadia, it was about 100 inches or more in the eastern districts.[5]

The Sundarban forests stretched across the sea-facing southern fringe of the three districts of 24 Parganas, Jessore and Barisal. Most of the forest areas that fell within the 24 Parganas district of western Bengal were affected by salinity as a result of the loss of headwaters from the Ganges, which had changed course towards East Bengal. Reclamation and cultivation in this part of the Sundarbans were, therefore, relatively less intensive. The government's decision to declare most of the 24 Parganas Sundarbans as a 'reserve' forest zone in the 1870s was informed more by this fact than by any visionary conservation ideals. Around the middle Sundarbans, lying in the Jessore district, the water was 'tolerably sweet'. In the third portion of the Sundarbans, in the district of Barisal, the Meghna river afforded a zone of sweeter water and a 'pleasant change' from the 'hot, dry and swampy atmosphere' of the western Sundarbans. In this region, few embankments were necessary to prevent the saline water of the sea from entering inland, since the strong river currents were able to contain the seawater. In this part of the Sundarbans, reclamation and cultivation were so extensive that in most places forests were cleared right up to the coast.[6] (Figure 3.1 shows the position of villages close to water bodies.)

Although productive exploitation of the agro-ecological zones of eastern Bengal had been under way from the pre-colonial period, the region seems to have been specially favoured by nature since the 1830s. The colossal changes in the river system, which saw the completion of the eastward journey of Ganga culminating in its meeting with Brahmaputra near Dhaka, had stabilized by the 1820s. A spell of earthquakes resulting in death and diversion of main waterways

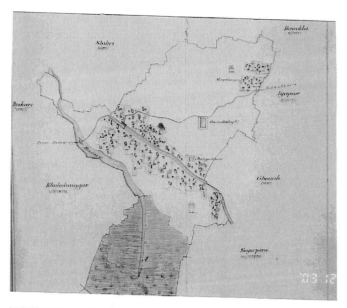

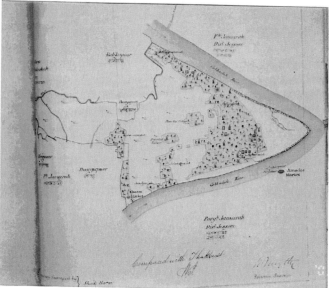

Figure 3.1 Sketch-maps of villages in proximity with water bodies in Khulna, 1840s

Source: National Archives of Bangladesh, Dhaka; Khuna Records, Bhalukha Thana.

along with major floods had also taken place over the last half of the eighteenth century. Even though these purely natural phenomena caused dislocations in economic and production conditions in Eastern India in general, East Bengal seems to have suffered relatively little from them.[7] By the 1830s, major tectonic unrest along the Himalaya seem to have subsided, the deltaic frontier being left exposed only to seasonal disasters such as tidal upsurges and cyclones.

The agro-ecological regime that evolved in the East Bengal frontier was 'admirably adapted to, and capable of, producing every kind of crop'.[8] Rice, in keeping with neighbouring south-east Asian monsoon territories, continued to be the primary crop in the region for both subsistence and trade. Rice had three broad varieties: *aman, aush* and *boro*. The aman, or winter rice, was sown between April and June and reaped between November and January. Grown mostly in depressions or marshes it was an impressive plant which grew 14 to 18 feet long and could sprout up to six inches per night and therefore was able to 'run a race against inundation, and beat it'.[9] The aush or summer rice (*Ashurvrihi*, the quick-growing) was sown, grown and reaped in about 100 to 120 days. It grew after the first showers of spring and was harvested in July or August. The boro variety grew mostly in marshes and was planted between January and March for harvesting by May. The boro too was self-sustaining, and provided that its roots were wet, the 'fiercest heat would never injure it'.[10] It was too tall to be absorbed by the water of the marshes. Gastrell saw a single stem of the boro variety upwards of 30 feet long in the great swamps common to Faridpur and Barisal. Whatever the predicaments of inundation, all varieties of rice appeared uniquely suited to the local ecology.[11] Jute as a commercial product owed its predominance in Bengal more to the region's soil regime than to the monsoon. Attempts were made to cultivate jute in West Africa and British Guiana with financial support from the Dundee manufacturers, and small crops were successfully grown in Japan and China, but Bengal raw jute enjoyed an effective monopoly.[12] Between the two principal varieties of jute in the Kolkata market the best was the one that grew in East Bengal; the other variety, generally known as deshi, was produced in the districts surrounding Calcutta. According to Hunter, the chief cause of the superiority of the East Bengal jute seemed to be the 'richness of the soil and the fructifying inundations'.[13]

The volume of jute exports jumped fortyfold between 1838–43 and 1868–73, and during the 1860s the value of raw jute exports increased from Rs. 4.1 million to Rs. 20.5 million. With only occasional and temporary slowdowns during the recession of the early 1870s, jute cultivation, production and export was in its heyday until the early twen-tieth century.[14] As far as rice was concerned, in the Chittagong division, for instance, there was a very large surplus which was exported from Chittagong port. Average annual exports in the late nineteenth century were about 1.5 million maunds, most of which was produced in the Chittagong division.[15] The largest share of the supply of rice for Kolkata was grown in the coastal districts, including the Sundarbans. In 1871–72 and 1872–73, between seven and eight million maunds poured in annu-ally to Kolkata port through the Eastern canals, which connected Kolkata to the Sundarbans, Barisal, parts of Jessore and the country around the Meghna river.[16] The amount represented almost 90 per cent of the total export of rice from Kolkata. Beside this, about 3 million maunds passed annually through Kolkata to various regions of India. The district of Barisal alone provided about three million maunds every year.[17] (Figure 3.2 shows a typical small ship plying a river in eastern Bengal.)

Trade and transport: bazaars and boats

'Rivers first create land, then fertilize it and finally distribute its produce' – the role of rivers in the Bengal Delta was thus summarized

Figure 3.2 Drawing by Robert Farren of a small ship sailing across an East Bengal river
Source: A.L. Clay, *Leaves From a Diary in Lower Bengal* (London, 1896).

by a hydrologist in the early twentieth century.[18] So far we have examined the first two functions of the river; the third and equally important role is that played by rivers and other water bodies in the distribution of the region's produce. Markets in Bengal were mostly riverside phenomena and were generally accessed by boat. Along with the three markets created by Henckell, the first official leaseholder in the Sundarbans, numerous markets grew as the reclamation continued. This was a welcome change in agrarian market relations since every producer-seller was subject to various tola or tax if he used a bazaar situated in a zamindari estate under the Permanent Settlement. Lal Behari Day observed that in a 'hat' (weekly or bi-weekly bazaar), the Brahmin would ask tola for puja and the phandidar would ask tola for 'safety' of the villagers. The zamindar would not take tola for the ground used for the hat, but would reimburse himself by taking a small quantity of the goods in which the sellers dealt. Day remarked that by adopting this method of remunerating himself, the zamindar got 'a hundred times more than he would have obtained if he had charged a fair rent for the ground'.[19] The creation of new markets by the grantees or leaseholders in the Sundarbans expanded the scope for cultivators to sell their produce without such barriers. The reasons for the development of an extended market network in the Sundarbans was that the Sundarbans itself was the site of production of a huge volume of export produce, notably rice, and that it also possessed the principal river routes to the port of Kolkata. These Sundarbans markets, including Chandkhali, Paikgacha, Surkhali, Gauramba, Rampal (Parikhali) and Morrellganj, were used not only by the peasants who lived in the territory of a particular ijara or lease, but also by peasants of adjoining and more distant areas under the Permanent Settlement. The peasants would bring their produce by boat to these markets in three to six hour pulls and sell it themselves. Robert Morrell, a leaseholder and founder of Morrellganj bazaar, informed the Indigo Commission[20] that he used to let the shopkeepers have their lands rent-free for six or seven years, and reduced the price of salt from eight to six pice per seer by procuring salt direct from the Board of Revenue. By doing this, he 'induced the ryots to come stealthily' to his bazaar, which gradually became the largest in the Sundarbans.[21] In Chandkhali, larger transactions took place once a week when ships and boats arrived from various directions laden with grain, buyers and sellers. People who traded in everyday necessities also came with their products to meet the demands of the 'thousand ryots' who had brought grain to market and would buy a week's store of daily necessities. From early morning on the day of the hat the river and the khal became alive with local crafts and boats, 'pushing in among each other and literally covering the face of the water'. Sales went on rapidly and the baparis and mahajans (traders and merchants respectively) filled their ships with the grain bought from the peasants. On average, 3000–4000 rupees worth of rice changed hands every hat day, when about 1500 boats of different sizes were brought up.[22] Besides the trade in these bazaars there was an immense traffic in goods by traders stationed all over the Sundarbans. Some of these traders had large ships and visited the reclaimed lands and filled their ships close to where the grain grew.[23]

Apart from the coastal and Sundarbans areas, trade and commerce further inland was facilitated by an intricate river network. While rice and various forest products were the principal trade items in the Sundarbans, jute and rice were generally traded further inland. Sirajganj and Narayanganj were two major marts, the former serving as the key centre of trans-shipment where the smaller boats, after navigating the upper river reaches in Assam, transferred their cargoes into larger boats which were able to withstand the more dangerous navigation of the lower deltaic rivers connecting to Kolkata. In the early 1880s, William Hunter estimated the total annual value of river trade in Brahmaputra to be £3 million sterling (about 55 million rupees).[24] In the mid-1870s, Richard Temple described the scene of trade and trans-shipment in Sirajganj in the following words:

> Boats of all sizes in thousands are moored and lashed together, thus constituting stages, almost roadways along which people can move to and fro. Tens of thousands of boatmen, workmen, and traders are congrated; this concourse induces villagers and tradesmen to bring supplies on board the boats; the merchants find it convenient to arrange their transactions on board also. Thus a floating city is actually formed on the river for several months in the year; on board of this vast flotilla, markets are held, goods disposed of, even rates of exchange settled, and transactions proceed as if on land ... the river-banks and temporary islands mid-stream change every year, therefore the floating city, while it keeps its name ... it has been not inaptly termed a town without houses.[25]

In other bazaars, such as those at Barisal and Nimtali, most of the export trade took place between November and March. During the rice season the bazaars presented a vivid picture as boats came from all parts of Bengal, especially from Kolkata. Beveridge noted that there was great demand for silver during the rice season and that the local currency was soon cleared by the presentation of currency notes and supply bills.[26] In all these transactions, boats were so central that it was believed the exemption of eastern Bengal from famines had something to do with boats. As was noted in a puthi, in the context of famine in Bihar and West Bengal in 1875–76:

> We have heard that in the North and in the West
> Numberless people are dying of starvation
> Boats do not ply in those countries
> Otherwise, would they die in this reign of the Queen? [27]

The waterways and the country boats remained extremely important elements in the marketing of deltaic produce, even though steamers and railways offered strong competition towards the end of the century. The loss of time incurred by boat travel was 'comparatively a trifling consideration' to most people when contrasted with other means of transport, because the cultivators had to pay high rates of freight for the steamers and numerous incidents of theft took place at railway stations. These considerations encouraged the use of boat transport to such an extent that in the late 1870s it was reported that the 'boat had taken so deep root in the land that no power of iron and fire will ever be strong enough to eradicate it'.[28] The government appeared to be ruthless in ensuring the free flow of boat trade. Westland noted that by the turn of the nineteenth century some zamindars from Jessore had set up about 18 toll houses along the fourteen-mile water passage from Chandkhali to collect illegal tolls from the trading boats that passed through their estates. The government responded by instituting corporal punishment, which was applied to the agents of zamindars as well as the zamindars themselves.[29] No wonder that although the railway began to penetrate East Bengal in the late nineteenth century, country boats continued to carry the majority of both rice and jute.[30] The persistence of indigenous transport here reflects the fact that the region's water system was relatively good, which also increased the importance of its economic role. Convenient access to the flourishing network of bazaars by the cultivators meant they could take proceeds home directly from their sale. In 1864, the price of a 400lb bale of jute was Rs 19.15. In 1894 it was Rs. 33.7 and in 1914 it rose to Rs. 80.[31] The steep rises in price were the result of a phenomenal increase in demand for jute in the USA and Europe in the second half of the nineteenth century. Over the course of the nineteenth century, increasing rice prices in the Bengal districts were also the result of increased demand, although this was in the context of scarcity and famine in other parts of India, such as in Bihar (1874), western, northern and southern India (1877–78) and Madras, Bombay and Bihar (1891–92). Bengal rice also had a widespread international market until it was replaced by Burmese rice in the 1910s.[32] Peasants living in the non-permanently settled areas, where lack of control by the landlords and the ecological advantages of production and transport offered direct and unobstructed marketing facilities, were able to benefit from these high prices.

Much of the literature on the process and impact of the commercialization of agriculture in colonial Bengal relates to a 'transition' period to colonialism. C.A. Bayly's argument against the formerly predominant notion of the total collapse of the Indian economy with the decline of the Mughal empire has influenced a number of works on Bengal that take a more nuanced approach to the problems of continuity and disruption.[33] Another important range of literature on Bengal agricultural development relates to the era of depression in the 1930s, leading to the famine of 1943. Most of these works are concerned with the role of 'colonial capital' in Bengal's agriculture. There is less focus on the social and economic impact of commercialization in East Bengal in the long nineteenth century. An understanding of the dual impact of the nineteenth-century's formative ecology and commercialization of agriculture is important to a proper perspective on agrarian and social decline in the late colonial period, an attempt we make later in this book. The following sections trace the degree of material and physical well-being and aspects of social formation.

Between Shangri-La and colonial dystopia: mapping the everyday economy

Although more recent research tends to support the contention of relative economic buoyancy in nineteenth-century East Bengal, there is no agreement on the actual duration of such a buoyant phase. Sugata Bose argues that the smallholding peasant economy of the region looked healthier than that of other parts of Bengal as late as the 1920s, at least in terms of the absence of gross income inequality, landless labourers and poverty.[34] Following Binay Chaudhuri, Willem van Schendel and Aminul Haque Faraizi believe that a process of 'depeasantization' was accelerating by the 1890s and resulted in the emergence and growth of numbers of vulnerable sharecroppers and landless labourers over the following century.[35] Schendel and Faraizi's point of departure is the Dufferin Report, a government report of 1888 on the condition of the ordinary people of eastern India. The report was produced in an era when radicals in Britain and the Indian nationalists were beginning to challenge the colonial government's claim of improvements in public welfare in India, hence much of the report was cast in excessively optimistic tones. In fact, one of the aims of Schendel and Faraizi was to challenge the report's suggestion that 'the peasantry of Eastern Bengal are about the most prosperous in the world'.[36] Nevertheless, the Dufferin Report was comprehensive and, despite some methodological weakness, it 'did provide a certain consistency and some comparability to survey findings from different parts of the country', as Schendel and Faraizi observe. Therefore, within the report itself it is possible to read some of the narratives against the grain as it offers an opportunity for comparison between different regions of eastern India.

Enquiry for the report was designed to collect facts in limited areas or villages, chosen as typical of the districts in which they were situated. In the selected villages (about 100 in number), information was sought with respect to the areas of the holding of each raiyat, the utensils that he used, the extent to which he was indebted, the value of the ornaments worn by the female members of his family, estimates of the produce of the land, the cost of cultivation, ordinary food and clothing, wages for labour and the degree to which the people were independent.[37] The whole population of a selected village was divided into the following eight categories for the purpose of retrieving specific information:

1. Cultivators, pure and simple.
2. People supported partly by agriculture and partly by wages of labour.
3. People supported partly by agriculture and partly by trade or profession.
4. Landless day-labourers available for hire.
5. Artisans, pure and simple.
6. People who subsisted only by trade or profession (except artisans, barbers, etc.)
7. Barbers, washermen, drummers, and other such classes who had no land.
8. Beggars.

In the report, the cultivators 'pure and simple' or the raiyats, who comprised the bulk of the population, were in the most advantageous position. They possessed land ranging from four to twenty acres and were able to sustain a family all year round and save money from the sale of rice or jute. They were reported to be 'well-fed, well-clothed and well-housed'. In the Chittagong division, comprising the districts of Comilla, Noakhali and Chittagong, one of the reasons for the relative material well-being of this category was that many of them were also traders in grain or cash crops. Especially in southern Comilla and Noakhali, the raiyat's extra income came from the sale of betel nut. Thus at a khas mahal in Noakhali some ordinary raiyats achieved a net profit of Rs. 16 per year in terms of the paddy alone, producing an average of 45 maunds per acre.[38] D.R. Lyall, the commissioner of Chittagong, believed that the average data for the 15 successive years from 1872 conclusively showed 'a very large balance in favour of prosperity'. He also noted that the people had the means to 'tide over bad years without extraneous help' and that they had the surplus to depend on.[39]

It was the commissioner of the Dhaka division, which comprised the districts of Mymensingh, Faridpur, Dhaka and Barisal, who reported that eastern Bengal peasants were about the 'most prosperous in the world'.[40] In the district of Barisal, the census of 1881 showed that about two-thirds of the whole population were of cultivator class. The average number of inhabitants of each house or family in the district was found to be 8.45. Based on the fact that there were about 1,371,629 people in an area of 2235 square miles of cultivated land in the district, the holding for each family was on an average close to nine acres. The average produce of paddy per acre was found to be 26 maunds which gave 13 maunds of rice. Therefore, if each cultivator employed his own labour, he and his family would obtain 117 maunds of rice a year. If each member of the family of eight used half a seer of rice a day,[41] yearly consumption would have been 35 maunds, leaving 81 maunds worth (at Rs. 2 a maund), Rs. 162, to pay rent and purchase necessities. The average rent and taxes for each holding was about Rs.45 and a large allowance for each family on account of clothes, salt, oil and so on was about Rs. 8 per month, or Rs. 96 per annum. Thus, at the end of the year the cultivator could have had at least Rs. 20 to spare, without taking into account the value of the straw or of the pulses grown in the cold weather, or of the valuable crop of betel nuts from the palms with which every house was surrounded.[42] (See Figure 3.3.) In the khas mahal areas of Chittagong, for instance, silver ornaments, golden noloks or nose-rings were commonplace. People attached so much importance to this piece of ornament that half of the country ballads were dedicated to the nose ring or to its 'fair owner'.[43] The idea of the peasants' prosperity appeared so convincing to

Figure 3.3   Drawing of a village in a Sundarban clearing. Pen and ink; worked
up from an earlier sketch of January 1839
Source: IOR, WD4360.

those who observed the peasantry from a distance that they noted that
the 'hard savings' of the peasantry remained buried in the ground, and
were spent only to meet legal expenses, if required.[44]

Members of the second category, raiyats holding less than 4 acres, generally supplemented their cultivation by selling their labour, although
many of them subsisted on very small holdings by concentrating on the
more valuable crops, such as jute. Lyall considered them 'quite well off,
if not better than class I' and certainly not suffering from any insufficiency of food. The members of the third group generally did not cultivate their lands themselves, but let it out on barga (sharecropping) basis
and received half the produce, with the cultivator keeping the other
half. In Nabipur Hussentala, a typical village in Pargana Bardakhat in
the Mymensingh district, 213 families consisting of 1035 persons were
divided into the eight prescribed classes. The persons shown in classes
3 and 6, who constituted 9.8 per cent of the population of the village,
were found to be in a 'prosperous' condition. It was reported that after
meeting all expenses, including those relating to marriage and other
social ceremonies, they had the 'power to lay by some money'.

Group 5 or 'artisans pure and simple' who had no land showed signs of mixed fortunes. It was generally found that while blacksmiths and carpenters could somehow manage life, weavers were living in the worst conditions. It must, however, be kept in mind that throughout the nineteenth century a large number of weavers became cultivators. The 1872 Census of Bengal shows that among the weaver class of the deltaic districts, only about 45 per cent were actually engaged in the weaving profession. The rest were assumed to have 'gone to the fields' (Appendix 4).[45] Some weavers were compelled to work in the Kolkata and Dhaka mills, but the rest, for instance those in Comilla, would not go anywhere else and would not do anything other than engaging in their ancestral professions. Some officials saw this 'caste prejudice' as the main reason for their comparatively poor economic condition.[46] But a more plausible explanation appears to be that the purchasing power of the average peasant was used for buying European cloth rather than cloth produced by local craftsmen, resulting in lower incomes for local weavers.

Members of group 7, the landless barbers, washermen, drummers, fishermen and sweepers, were generally worse off than those who had land. The colonial officials again assigned their economic decline to their caste-induced attitude of sticking stolidly to their traditional profession. However, this explanation ignored both the interaction that took place between these members of the 'service sectors' and fluctuations in rural productivity. Any temporary decline in agricultural production as a result of natural disaster caused members of this group to feel the pinch. If a cultivator's family had less than an ordinarily available amount of rice in a particular year, he would possibly have spent less in securing service from washermen, sweepers or drummers, or indeed on buying fish.[47] Regarding the last group, beggars, the report considered them as worse off than the members of group 7, although in some respects the beggars appeared relatively better off. In Dhaka, they were 'not very numerous' and managed to support themselves at a level of 'tolerable comfort'.[48] In the Barisal district beggars were estimated to be not more than 2000 in all, but all of them were reportedly old prostitutes who had turned Vaishnavites and who, as a rule, found begging so profitable that on their death they left a 'very handsome legacy to their gurus'.[49]

The fourth category, that of landless day-labourers, provided perhaps the most important index of economic conditions of the people.[50] One official reported that there were no landless labourers in eastern Bengal. But an analysis of all the reports sent to the enquiry committee suggests that this was not entirely true. The report suggests that in some khas mahals, there were indeed no landless labourers. For instance, in Nalchira of Noakhali every man possessed a share in a piece of land sufficient to prevent him from working as a day labourer for daily wages.[51] But in the villages examined in Salkwa khas mahal in Chittagong, 290 out of 887 families were classified as labourers. In Cox's Bazaar the number was about one fourth or 1724 out of 4405. Even if we deduct from this pool of labourers those who were only taking recourse to labour after exhausting their own land resource (group 2), the number of day-labourers reliant on wages would not be insignificant.[52] Indeed the village surveys revealed that 26 per cent of the rural Bengal population had labour as their only or main occupation.[53]

Having said that, we need to get a clearer idea of what 'landlessness' meant in eastern Bengal in the nineteenth century. It seems that 'landlessness' was not synonymous with having no land at all. A 'landless' person might not have had enough land on which to produce rice or jute on commercial basis, but he might nevertheless have possessed some land around his bastu or home. For instance, it was reported that in the Comilla district all residents – including beggars – had homestead lands amounting to at least one-third of an acre.[54] But for a landless wage labourer, it was the wage rate that mattered most. In the Chittagong division, for instance, there were a very large number of cultivators working as agricultural labourers, and they went as far as Arakan for labour, returning with between Rs. 20 and Rs. 40 cash after two months' work. In addition, the people of Chittagong and Noakhali also supplemented their incomes by working at sea as lascars and firemen, and by acting as traders in Myanmar. In Chittagong, a large number of men also went each year to the government kheddas on high pay. Besides which, the people of Chittagong and Comilla also traded in hill products.[55]

In other parts of East Bengal, where sources of external income were not as flexible as in the districts of the Chittagong division, landless labourers who depended mostly on the sale of labour in the rice and jute fields were probably more numerous. In the Barisal district, for instance, a labourer received 4–6 annas a day. In Mymensingh, a labourer was generally paid Rs. 4–5 per month with food and sometimes with food and lodging. In Faridpur, whereas a raiyat earned Rs. 9–20 per month, a labourer earned about Rs. 4–9 per month. In most cases, labourers also supplemented their earnings by fishing.[56] It may also be noted that those who were considered as wage labourers in the Sundarbans during rice harvest were actually migrants from further inland, who after reaping their own crops, came to the southern region to earn additional income.[57] There were instances of people migrating seasonally to khas mahals to cultivate paddy. The migration took place not because the cultivators were always aspirant for occupancy rights or because they were desperately looking for subsistence, but because they sought additional profits. In some khas mahals of Pabna, people from the Natore subdivision came for cultivation of aman paddy. During the rains they went home, returning in the month of *Agrahayan* (in the autumn). If the crop was good, they paid rent; otherwise they went away, leaving the standing crops in the field.[58]

The authors of the Dufferin Report were aware of the rhetoric they were trying to promote, hence they often attempted to support their findings by reference to non-official sources. One report quoted Shambhu Chandra Mukerjea, a journalist who, while passing through Comilla in 1878, wrote that he was infinitely delighted to observe the evidence of 'comfort and comparative civilization in the peasantry' of this region. He noted that they were well protected from the cold by European clothes. The women all had costly ornaments, 'if less heavy and numerous than their mothers and grand mothers could boast of'. He found shell bracelets few and far between, he found no brass any-where, but silver 'clearly predominate[d]': on the banks of the Meghna he found 'all the women display silver – some in profusion'. He considered that it was all due to jute.[59]

How did the condition of the landless labourers in the Bengal Delta compare to that of eastern India in general? On the basis of the Dufferin Report, Schendel and Faraizi suggest that whereas the proportion of labourers in West Bengal districts ranged from 37–43 per cent, in East Bengal districts it ranged from 4–23 per cent.[60] The report observed, 'in the extreme east, where labour was most scarce, wages were highest, and the rate decreased almost proportionately as the survey travelled towards the west'. In the districts of Hugli and Burdwan in the Burdwan division, poorer cultivators and field labourers formed about 75 per cent of the population, which also appeared to be weakened by diseases such as fever and leprosy. In Burdwan, and to some extent Hugli, wages were close to those of eastern Bengal, but the scarcity of labour here was the result of high mortality rates following malaria epidemics. In these districts, though the people were physically weakened by sickness, and their labour was therefore less efficient, they found no difficulty in maintaining themselves, because their numbers were 'not permitted to rise to the point at which they would press unduly on the resources of the country'.[61] Symptoms of 'chronic poverty' among cultivators in general and labourers in particular were clearer in Bankura and Midnapur.

Here the daily wage rate appeared to be three annas a day, with employment being locally available for six months of the year only. It was reported that about 10 per cent of the population had only one meal a day. In the Bihar districts of Patna, Gaya, Muzaffarpur, Darbhanga, Sarun, Bhagalpur and Mongyr, while the upper and middle classes were prosperous, the ordinary labourers, and the smaller cultivators, amounting to some 40 per cent of the population, were much worse off than the corresponding classes in Bengal.[62] Daily wages in these districts were 'nowhere estimated at more than two annas' and in many cases did not exceed five or six pice a day. Rents were also comparatively high. What cost three to four pice in Bihar, cost six annas in eastern Bengal.[63]

Sources other than the Dufferin Report seem to lend support to the impression of the relative social and economic dynamism of East Bengal. On the question of wages, Richard Temple had observed a decade earlier, in 1876, that in general it was one to two annas a day in Behar, two annas in Orissa, three annas in northern Bengal, four annas in central Bengal and five annas in eastern Bengal.[64] A witness to the Indian Famine Commission of 1898 remarked that no man in Bengal would work in his own village if he could help himself and that 'he will move east' where there was a 'very small indigenous labouring population'. There he would get higher wages though the price of food was the same.[65] More recent research, based on detailed examination of district-level wage and food price data in the late nineteenth century, seems to corroborate these scenarios. William Collins suggests that in terms of real wages, East Bengal compared favourably not only with its neighbours in eastern India, but also to places such as Bombay, which had overall relatively higher wage rates. In terms of inter-provincial comparison, districts in Burma, East Bengal and Bengal had the largest product increases between 1873 and 1906 whereas those in Bombay, Punjab and the central provinces had the smallest product price increases. Collins believed this reflected the fact that rice and jute prices rose considerably more than wheat or cotton prices in the late nineteenth century.[66] The fact that the East Bengal districts had cornered the global market in jute would have given the region still greater advantages over Bengal and Burma. In examining the links between world commodity market integration and poverty in colonial India, Tirthankar Roy argues that the nineteenth-century period of globalization had a broadly positive impact on rural economies and incomes, but that the level of well-being was shaped and constrained by 'local resource conditions and patterns of labour mobility'. Roy's example of Bengal suggest that until the last quarter of the nineteenth century the daily wage gradually increased and that inflation was much less than was the case in the first quarter of the twentieth century.[67] If the local resource conditions and patterns of labour mobility are taken into account, much of the positive impact of this must have been felt in East Bengal.

As far as the question of indebtedness among the peasantry is concerned, the situation in East Bengal should be viewed in the light of the reclamation process. In the debates on labour migration from the economically backward regions to the regions where wasteland abounded, the Bengal government suggested that inland immigration was better than immigration to foreign countries. This was based on the general assumption that in the context of the ownership of tenants' rights in a fertile soil and with the rising price of food, the reclaiming cultivators theoretically ought to be on the 'high road of competence if not wealth'. But in some areas of Bihar, such as Hazaribagh and Bhagalpur, the reclamation process contributed to an alarming state of indebtedness as it depended on loans extended by the zamindars and mahajans. In East Bengal, where reclamation was more extensive than in Bihar, the situation was different. For instance, in Chittagong and Noakhali, the villagers did not generally borrow money at all. Those who borrowed did so for specific purposes, for example, to fund a lawsuit, meet marriage expenses, or pay nazr to get a lease. The loan was temporary, and was 'always paid off after the next harvest'. The sources of borrowing were not bankers or professional money-lenders but fellow villagers. It was reported that whereas the profits from agriculture had increased, these profits had gone into the pockets of the raiyats, and the most indebted class was the landlords, who were debarred (by the government) 'in every possible way from getting their fair share in the increased value of produce'.[68] In Pabna, debt amounted, on average, to about two months' income. The debt situation further improved after the stoppage of indigo cultivation.[69] It was estimated that probably one half of the raiyats were 'entirely free from debt'. In Rajshahi and Sirajganj, the situation was 'simply better'.[70] In Barisal, the raiyat had 'no conception of the squalor and poverty of the Behar, Burdwan or Hooghly ryot'. Here a man with tenure could borrow more cheaply than a tenant-at-will, since his security was better. But there were 'almost no tenants-at-will in this district'. Whereas in the Mymensingh district, 'Mahajan proper' hardly had a place, in Comilla, village mahajan was 'almost unknown'.[71] In general, whereas debt was 'worse in Behar, somewhat considerable in Central and Western Bengal and Orissa, [it was] less decidedly in Eastern and Northern Bengal' and it was 'altogether disappearing in parts of Eastern Bengal and Northern Bengal'.[72] It was no wonder that there was evidence of 'substantial net flow of migrants from Bihar and Orissa to Eastern Bengal'.[73]

We may close this section by quoting Ramesh Dutta, a vocal nineteenth-century nationalist critic of the colonial government. Recollecting his personal visit to East Bengal in 1876, more than a decade before the publication of the Dufferin Report and at a time of famine in Bengal, Dutta noted:

> The peasantry in those parts [Eastern Bengal] paid light rents, and were therefore prosperous in ordinary times. With the providence and frugality which are habitual to the Indian cultivator, they had saved in previous years. In the year of distress they bought shiploads of rice out of their own savings. There was no general famine, and no large relief operations were needed. I watched with satisfaction the resourcefulness and the self-help of a prosperous peasantry. If the cultivators of India generally were as prosperous as in Eastern Bengal, famines would be rare in India, even in years of bad harvests.[74]

The state of health and disease

The Dufferin Report, however interpreted, concentrated on the *economic* prosperity of eastern Bengal. Some contemporaries and later historians were well aware that ecology and economic structure working together affected the well-being and health of people in a much wider sense. This section examines some of the issues relating to health and diseases.

Although most theories which attribute disease to climate now sound medically suspect, a discussion of health and disease in Bengal requires an attention to indigenous knowledge and popular attitudes towards 'tropical' climate and landscape. Writing in the 1830s William Twining found that people generally exposed to extreme heat in Bengal were liable to apoplexy, paralysis, inflammatory fever and sudden attacks of cholera.[75] The general attitude of the Bengali people to Bhadra (August–September), a very hot and humid season, provides an example of the popular response to the region's climate. Twining observed that the month of Bhadra produced 'extreme languor, depression of spirit, and exhaustion of bodily strength as well as mental energy'. During this period, animals languished and became sick, and consequently meat was of 'indifferent quality'. It was estimated that a fourth of the total number of annual deaths occurred in Bhadra. Twining noted that 'little superstitious people' considered this month not only unhealthy but unlucky and believed that those undertaking a journey in this month were liable to 'lose their lives, or to have their health permanently impaired'. Marriages did not take place among the Hindus and women did not visit their relatives. Twining remarked that many of the popular attitudes towards this month were the result of long observation.[76]

Although popular perceptions of the warm climate were identical across Bengal, in East Bengal, which experienced the first spell of monsoon each year, the average temperature was cooler than in other parts of Bengal or the northern Gangetic plains. Here the average temperature was 15 degrees Fahrenheit lower than in Allahabad, Kanpur and Delhi and at least 9 degrees Fahrenheit lower than in the West Bengal districts of Midnapur, Burdwan and Birbhum during the hottest months of the summer. On average, the mean annual temperature in eastern Bengal was 3 to 4 degrees Fahrenheit lower than that of western Bengal, although both were situated along the same parallel of latitude.[77] However, the lower mean temperature in eastern Bengal did not mean that it was out of the reach of tropical heat and humidity and the Bengalis in East Bengal shared the same concern about the hot climate as fellow Bengalis in other regions. Those who could afford it avoided 'any exertion', consumed less food, and more fruit. In the afternoons, they drank the fluid contained in the unripe cocoa-nut or a simple *sherbet*, or some sugar and water, which was thought to be especially cooling. They also occasionally drank the juice of the leaves of *Nalta Pat* or jute (*Corchorus olitorius*) in the mornings as it was considered to have a 'cooling and mild tonic effect' as well as digestive powers. Twining observed that though adapted by nature to bear the climate, the Bengalis took 'more care to moderate the effects of the hot-season than Europeans – especially in their light clothing, abstemious food, and tranquil habits'.[78]

Despite all these precautions, however, western Bengal and to some extent northern Bengal, suffered more disease and deaths from disease than eastern Bengal, where, for instance, the problem of malaria in the late nineteenth and early twentieth century was 'only a fraction as severe as that of West Bengal'.[79] The factors that contributed in some way to better conditions of health in eastern Bengal than in other areas, therefore, appear not to be related specifically to heat and humidity. One such factor was the style of housing, particularly in regard to ventilation. Twining observed that poorly ventilated living spaces caused a number of diseases, including cholera.[80] Kanny Loll Dey, a Bengali sociologist, observed in the 1860s that continuous changes in the airflow were needed in order to keep it 'in its necessary state of purity'. Following the estimate of a European doctor, Dey thought that the change must amount to at least 2000 cubic feet per hour per head for persons in good health, and not less than 3000 or 4000 cubic feet or more for sick persons. But he observed that a room in western Bengal was like a 'hermetically closed box'. In the masonry buildings in which better-off people lived, the most spacious and airy parts of the building were reserved for social functions such as pujas, and in these areas male members of the family predominated. On the other hand, the zenana, where the women of the family lived all day and night, received hardly any air or light. Day noted that when the male members entered the zenana for rest or a night's sleep, they joined their female members in the same 'hermetically sealed' room. Therefore, sleep, 'nature's sweet restorer', brought rather 'lassitude and enervation'.[81] In the accommodation of ordinary people, who could not afford masonry buildings, house walls were made of mud and there was very little ventilation, except through the door.[82] Compared to this, in eastern Bengal, the accommodation of the ordinary peasantry was mostly built of slit bamboo which contained 'a superabundance' of air.[83]

In the early nineteenth century, John M'Clelland observed that most of the soil of eastern Bengal consisted of grey sand, which, saturated with moisture retained by the clay on which it rested, was rendered rich and fertile during cultivation. When neglected, it degenerated into a poor sandy or heavy clayey soil, which was soon overrun with coarse grasses and other indigenous vegetation, which, once established, was difficult to eradicate.[84] It was, therefore, observed that in the eastern Bengal alluvial lands where reclamation and cultivation took place extensively, diseases were proportionately less prevalent than in other areas of Bengal where reclamation and cultivation were less intensive.[85] M'Clelland also observed that in the areas composed of lighter sedimentary soil, such as the Sundarbans, fever did not set in 'until after the first fall of the rain', when the malaria had immediate effect. It lost its effect for a time when the rains had set in and reappeared when the rainfall was interrupted towards the close of the rainy season. As in the eastern Sundarbans, districts composed of sedimentary deposits became safe after November. On the other hand, districts composed of laterite or heavy clay, mostly found in western Bengal, took a longer time to dry after the rains had subsided and were not 'safe to enter until the middle of January'.[86]

In the course of the nineteenth century the expansion of reclamation to the sea in certain areas made the coastal inhabitants of eastern Bengal more vulnerable to cyclones and tidal upsurges. As early as the 1860s, a colonial surveyor, Gastrell, warned that if reclamation continued at the speed he saw at his time, cyclones would start to claim more lives, since the reclamation had reduced the Sundarbans forests which had acted as a protective 'belt' between the sea and land. Gastrell's prophecy came true in 1876 when a major cyclone hit the coastal region with devastating consequences. The death toll included not only the victims of the cyclone, but also those of the cholera epidemic and other diseases that followed. What is remarkable, however, is the formative demographic response of the East Bengal people to the impact of the cyclones. The cyclone decreased the population of Noakhali district by 2.3 per cent on average between 1872 and 1881. Some places, like the islands of Hatiya where a quarter of the population was lost and Sandwhip, which lost a sixth of its inhabitants, were proportionally much harder hit. During the next ten years, however, there was a rapid increase in population and the census of 1891 showed an increase of 23 per cent which was wholly due to natural growth. In Barisal, the cyclone drowned about 74,000 persons and the cholera epidemic that followed killed nearly 50,000 more; yet during the next decade the district added 13.3 per cent to its population.[87] This was, as the census report of 1901 commented, a remarkable instance of the way in which a community could recover from the effect of a catastrophe of this kind without any assistance from outside.[88]

What made this possible? There are debates about the causes of remarkable population growth during the past two centuries in the Bengal Delta. But at least three processes must have been at work in the nineteenth century. First, the prospect of using more and more family labour in the land reclamation process; second, huge migrations to the more fertile deltaic regions; third, relatively better health and nutrition. In what follows I will examine the last aspect in a little detail, since the other two aspects have been discussed above.

In the Dufferin Report it was revealed that in Dhaka, agricultural labourers, who were relatively less privileged than the landed cultivators, went fishing when not otherwise occupied.[89] In Barisal, the collector recalled that he had never heard of anyone who did not actually cook food twice a day. In Noakhali no one went to work until he had a good meal. The day's work, too, was very short and easy in Noakhali which probably made the people 'stronger'. Along with this, the ordinary cultivators bought plenty of fish, meat and poultry items, and lived 'generously'.[90] In Noakhali, where agricultural labourers 'did not exist as a separate class', the poorer raiyats (group 2 above, or people supported partly by agriculture and partly by the wages of labour) worked for others during the two agricultural seasons, lasting about four months in all. They earned 4 to 5 annas for a day's work, besides a rich mid-day meal. This was the custom of the district and men would not work if they were not well-fed on hire.[91] In Baliakandi of Comilla, members of the second group used to sell part and consume part of their dairy and poultry products. They did not generally purchase fish which were 'almost always caught' in the neighbouring beels, rivers, canals and tanks. Neither did they purchase pulses, chilies, tamarind, vegetables and fruits, which were grown on their bastu.[92] When food scarcity became unusually acute due to excessive flooding in the Faridpur district in 1870, people were helped by the 'abundance of fish' and also by date trees.[93] In a similar situation, in the western district of the 24 Parganas people ate grain reserved for sowing, the leaves and roots of the plantain tree, the sajina, tamarind or other trees.[94] In 1868 it was reported that owing to good crops and high prices, substantial raiyats of East Bengal indulged more generally in 'animal food'. This explained the shortage of poultry products in the moffussil where these products were so scarce that prices had more than doubled, and were as high as those in Kolkata.[95]

A far more pessimistic picture of food availability was given by the collector of Comilla. According to him, in two typical villages of this district, groups 3 and 6 never suffered from insufficiency of food. Individuals in group 2 were not as respectable as those in group 1, but they were better off, and both classes suffered from insufficiency of food only after a succession of bad harvests. A single bad season would not greatly affect the supply of food as far as these two classes were concerned, but it would create suffering among members of class 4 or landless labourers. Groups 5, 7 and 8 were practically 'always in straitened circumstances'. The collector, therefore, believed that altogether at least 10 per cent of the population suffered from chronic insufficiency of food. It was, however, reported by Lyall, the commissioner of the Chittagong Division, that this picture of limited access to food was drawn from two villages which had suffered severely from inundation for three consecutive years, meaning that the figures showed the state of the people of that area 'at their very worst after a succession of bad years'. At the same time, the collector of Comilla himself noted that by 'insufficiency of food', he meant 'anything less than the usual three meals a day, the first eaten cold, and the other two hot'. The collector of Comilla remarked that these meals were generally supplemented by the personal efforts of individual members of the family, who went fishing in the neighbouring beels, kept poultry and livestock, and made some profit by selling livestock as well as eggs and milk.[96] A labourer who had no land or sufficient dairy or poultry got work in the fields for about eight months in a year on an average pay of Rs. 3 per month, besides securing food for himself. His annual earning, therefore, was Rs. 24 in cash, besides food for himself, with which he could maintain a wife and a child. His wife would independently husk grain for wages, and a boy, if 10 years old or more would tend to the cattle of others for Rs. 1 per month, in addition to food. The manager of Sarail estate reported that the labourers generally had a 'robust and healthy appearance', and that they did not suffer from 'under-feeding or want of due nourishment'.[97]

Jatek-gathan: aspects of social inclusion and tensions

While the new land reclaimed from chars and forests remained the principal theatre of agrarian and economic relations, it did not necessarily mean that social formation was taking place entirely 'outside' the territories of the Permanent Settlement. The pioneers of the reclamation process mainly came from the permanently settled areas, where economic vulnerability and a semi-feudal agrarian relationship pushed the cultivators to seek alternative options. The process of reclamation, therefore, did not denote a mechanical mode of cutting forest or improving char lands for cultivation. The process always retained the essential imprints of the society from which the reclaimers and settlers came. In most cases, reclamation was taken up in areas far beyond the vicinity of a settled tract or village. But even in the cases where distant migration took place societal relations were remarkably reproduced. For instance, in Barisal Sundarbans, the gradual extension of the population southward was brought about by means of the establishment of daula bari or a second home. A settled raiyat from a cultivated village put up a daula bari at a reclaimable or partly reclaimed tract, which he occupied during the cultivating season only. As the new tenancy grew in value, the daula bari tended to become a more permanent homestead to which the family, or more generally a part of it, migrated.[98]

Among those who reclaimed wastelands, nine out of ten cultivated it with their own hands, though they might have employed others to assist them.[99] This collective process of reclamation and settlement bonded the reclaiming tenants together in a spirit of equity, if not equality. They developed a mutual dependency as well as collective ways of doing things. It was the custom of the raiyats to offer one another mutual assistance and recourse to hired labour was unusual.[100] As the old aristocratic Muslim families and upper caste Hindus were not involved in the actual reclamation process, differentiation did not become a dominant feature of society. Beveridge was surprised by the dearth of aristocratic Muslim families in Barisal, which was full of Muslims.[101] The absence of north Bengal-type jotedars in the region may also be explained in this context. As new lands were being created or reclaimed from forests, tenants were at considerable liberty to settle on terms best suited to their demands. If they were not offered favourable terms, they would leave to seek better deals elsewhere. This was not conducive to the stability of the land market nor to the interests of the jotedars, who controlled these markets.[102] The magistrate of Noakhali, for instance, reported that there were very few wealthy individuals in the district, but on the other hand there were no paupers and the possession of wealth was 'widely diffused among all classes of the community'. He attributed this 'fortunate condition' to the system of land tenures which ensured the possession of a small plot of land or interest by 'almost every individual'.[103]

Among specific social groups, the Badyas were a migratory tribe who used to live on boats, and rarely set foot on land 'save for the purpose of theft or to sell their rude manufacturers [sic]'. In the course of the nineteenth century they formed colonies of the ordinary agricultural type, gradually merging into the vast mass of the peasant population. Though assimilating themselves in religion and cultural practices to the 'low-class' Muslims, a remarkable instance of their cultural survival was to be seen in their villages, which appeared as 'a congeries of mat huts, of a shape identical with that of the cabins on their floating homes'.[104] When the indigo industry was closed in the district of Sirajganj, another tribal group, called Buna, who had served as coolies at the indigo factories, were encouraged by the Taras zamindars to settle on the wastelands for the purpose of reclamation. Following this, the Bunas induced more of their kin from their homes in the districts of Burdwan, Birbhum, Bankura and Purulia to settle in this region, which included both zamindary estates and khas mahals. The Bunas, who were settled at moderate rates of rent, eventually appeared to be 'not anxious to return to their native homes'.[105]

The Namasudras, numbering about 1,860,000 in 1901, were the second largest Hindu caste in Bengal and the largest group of Hindu cultivators in eastern Bengal. They exemplify the linkage between the process of reclamation of wasteland and upward social mobility in the nineteenth-century Bengal Delta. The Namasudras did not exist as a distinct caste group, being considered an inferior caste loosely described as 'Chandals'. In the course of the nineteenth century most Chandals proved extraordinarily responsive to the opportunity of engaging in agriculture within the general ecological regime of the delta, which offered fertile land, higher wages for agricultural labour and profits from the commercialization of agriculture. The highly fluid environment of the marshy tracts where they lived and which they reclaimed for cultivation also often 'diluted the intensity of oppression by the dominant classes'.[106] Towards the end of the nineteenth century, the Chandals tended to identify themselves as Namasudra with a view to mobilizing along a distinct caste category. This transformation from Chandal, a rather nebulous cultural identity, to Namasudra, signifying association to higher castes of Sudras, was reasonably mediated by the status they gained through agriculture, which offered them a settled and prosperous life. Whereas at the beginning of the nineteenth century they had maintained an 'amphibious existence', earning their livelihood primarily through boating and fishing, by 1911 about 78 per cent of the Namasudras were reportedly connected with agriculture.[107]

The opportunity and fruits of reclamation of forests and chars and the extension of cultivation were comprehensively appropriated by the Muslim cultivators who formed the majority of the population – between 56 and 67 per cent in 1872. As Richard Eaton has shown, throughout the Middle Ages, reclamation proved to be the most important factor for the expansion of the Muslim population of the East Bengal deltaic frontier lands.[108] This evidently persisted throughout the early colonial period, and was subtly expressed in a popular ballad which narrates the movement of two brothers, Ghazi and Kalu, who started a journey saying 'bismillah' (in the name of Allah) and after travelling through many countries arrived in Bengal and at last settled in the Sundarbans where all tigers became their disciples. They stayed there for 'seven years'.[109] In sponsoring the publication of a puthi in 1837, the publisher said that his grandfather, who was from Mymensingh, migrated to Char Palash village where he built a home after cutting the jungle (*jungle katia teni bari banailo*). The puthi writer himself had a different story to tell. The writer, Siddik Ali, was an ordinary man and was too poor to make a journey. At one point, he began to learn about Islam and after finding an opportunity came to eastern Bengal and converted to Islam. It is not clear whether his migration to eastern Bengal was economic in nature or due to a genuine desire to lead a sacred life; but it was clear from his story that East Bengal was his chosen destination.[110]

The Muslim peasantry, like other social groups, were quick to avail themselves of the new agricultural opportunities. This was accompanied by an urge for self-esteem and honour. A collector of Noakhali reported that the Muslim cultivators were a thrifty class who seldom spent money on passing enjoyment and that it was the 'chief ambition of a Muhammadan rayat to save enough to buy a small estate, which will give him independence and position among his neighbours'.[111] Muzaffar Ahmad, one of the founders of the Communist Party of India, was born in the 1880s on the island of Sandwip. He noted that among the Muslims the title thakur (Tagore) was a symbol of some rank and the family of his grandfather had some claim to it, however small. He noted that in a small island like Sandwip, a family with even a touzi or two was considered aristocratic. His father was a mukhtear who was born in 1827 and he pointed out that nearly all his father's professional colleagues had become owners of a good amount of landed property. He also mentioned how many of his own friends were aspiring to a middle-class position in Sandwip and other neighbouring islands (such as Hatia).[112] This indicates that in those remote islands of the Bay of Bengal three consecutive generations went through a process of upward mobility, financially, socially or politically. A process of upward mobility paved the way for differentiation within the society. A puthi read:

> When a man becomes rich he does not care for anybody
> He considers himself above everyone else
> He has wealth but does not donate
> But when a *Hakim* visits his house, welcomes him heartily
> When [*ordinary*] visitors come he says there is no food at the
>   house
> When a *salar sammondhi* (distant but materially important guest)
>   arrives
> He kills pigeons [to entertain].[113]

Such social differences widened in the context of deteriorating agrarian relations as became apparent by the early twentieth century. But in the course of the nineteenth century, as long as ecological and social conditions remained favourable to the ordinary peasantry, society remained largely undifferentiated in an economic sense. Within these trends it was not surprising that a puthi written in the 1870s talked about jatek-gathan or 'nation formation'. The author of the puthi observed that the whole universe (jagat) was created because of the prophet of Islam while the nation (jati) was also created because of the prophet. It is not clear whether the author was speaking simply of the ummah of the prophet, or if he was making a point in the context of the emergence of a new society in eastern Bengal; but the term 'jatek-gathan' appeared particularly reflective of a mobile and vibrant society in which the writer apparently nurtured his sensibilities.[114] Although one does not miss the evident propensity to some 'Islamic' forms of identity, the nineteenth century also provided an interesting case for the hybrid, less purist forms of Islam.[115] Therefore, contestations between reformist and purist Islam persisted in many remarkable ways. While such contestations are documented in contemporary vernacular literature, one wonders about the relative lack of communal violence between Muslims and non-Muslims, at least in comparison to the extremely violent relationship that dominated the early twentieth century and culminated in the Calcutta and Noakhali riots and ultimately the partition of India and Bengal. It may not be an exaggeration to suggest that until the Swadeshi movement in the early twentieth century, tensions were more visible at intra-communal level than at inter-communal level

From the narratives of nineteenth-century agrarian society in East Bengal, several points emerge that we can connect with more recent historiographical debates. First, peasants in East Bengal responded 'rationally' to the opportunity offered by the agro-ecological resources of the area and the colonial state's relative weakness in regard to wilderness from the turn of the nineteenth century onward. They were keen to secure occupancy rights, using this entitlement to exploit the land to productive purposes and responding intelligently to the critical dynamics of the market.[116] The ecological regime of East Bengal was significant also in the way it enabled the peasantry to modify their production behaviour according to shifts in domestic and global commodity markets, at least until about 1900 when the connection between disturbances in agrarian life and uncertainty in the world commodity markets was becoming firmly established. For instance, whereas the East Bengal peasant could switch from jute to rice during the 1870s jute slump, the Bihar peasants could not abandon sugar and indigo.[117] While this flexibility was conditioned by the way East Bengal peasants, unlike the peasants in Bihar, were relatively self-sufficient and independent of the intermediary interests of the zamindars, it was also the deltaic ecology – its suitability for growing both rice and jute – that facilitated such flexibility. As this chapter has shown, the period of relative economic buoyancy had a lasting impact on the way in which the region's society was shaped and flourished in the course of the nineteenth century, illustrated by entitlement to food, better health and attempts to institute a religio-moral regime.

While all these can help to explain why East Bengal was able to escape the 'Victorian holocaust',[118] one cannot but ask why the region did not escape the wave of peasant resistance that affected nineteenth-century South Asia. Eric Stokes has argued that in later colonial India peasant resistance was launched both by richer peasants, especially in 'high farming' areas, and by pauperized peasants, the former aspiring to a more authoritative grip on agrarian relations and the latter venting their anger against injustice and accumulation.[119] Many subaltern scholars have tended to look at the nature of peasant resistance more in the context of communal consciousness than class. How do we locate peasant resistance in East Bengal within these debates? The next chapter will examine this question through an exploration of the history of the Faraizi movement.

4

The Political Ecology of the Peasant: the Faraizi Movement between Revolution and Passive Resistance

Bankim Chandra Chatterjee, the famous nineteenth-century Bengali polemicist, contended that the Bengalis lacked the physical as well as the mental stamina necessary to rise up against British colonial power in India. But he also argued that even if the Bengalis were weak and unfit, they could still emerge politically successful by employing inner strength of mind, or bahubal, which consisted of the qualities of initiative, unity, courage and perseverance.[1] Bankim proved himself correct. As Ranajit Guha has shown, the vigorous anti-colonial movement of the bhadralok in the early twentieth century reflected the fact that the Bengali nation had at last acquired the quality of bahubal.[2] However, the recent peasantization of the historiography of colonial Bengal raises issues with the categorical limit of Bankim's bahubal. The restoration of peasant agency tells that the peasantry already possessed the chemistry of bahubal for which Bankim was waiting in the emerging nationalist frame. This was poignantly represented in the Faraizi movement, which forms the subject matter of this chapter.

But why would the Faraizi peasants embark on a tumultuous career of resistance at a time when issues such as landlessness or acute hunger – as we have seen in the previous chapters – did not loom large? It is argued here that while the reclamation of new lands and forests was successfully linked to the commercialization of agriculture, leading to social and economic mobility, the connection was not automatic. Various agencies of the colonial state, such as investors in land reclamation projects and zamindars, persistently attempted to take a share of the productive value of ecological resources by restricting the peasants' access to land as well as resisting the accrual of occupancy rights and increasing the land tax. In other words, the space of formative social change which was created at the interaction between the global commodity market and the ecological resources was also the space of Faraizi resistance, which was launched when the sources of their relative well-being, such as reclaimed lands in the chars and forests, were at stake. This chapter, then, presents an ecological interpretation of the Faraizi movement.

A peasant's swaraj? A profile of the Faraizi movement

The Faraizi movement was started by Haji Shariatullah around 1830 (Figure 4.1). Born in the deltaic district of Faridpur in 1781, Shariatullah studied Arabic literature, Islamic jurisprudence and Sufism in Makkah and al-Azhar University in Cairo. On returning home in the late 1820s, he started a campaign within Bengali Muslim society to eliminate various traditional practices (riwaj) that contradicted the teachings of the Quran.[3] These riwaj included the worshipping of the shrines of pirs or saints, participation in the ratha yatra, the myth-inspired religious processions of the Hindu community, the planting of banana trees around the home on the occasion of the first menstruation of a girl and so forth.[4] These practices, often described as shirks (idolatry) and bid'as (unlawful innovations), were the objects of Shariatuallah's condemnation. But his attempt to cleanse the Muslim community from the 'syncretic indulgences' of rural Bengal led to indirect confrontation with the zamindars. Shariatullah interpreted the payment of unofficial taxes for the celebration of pujas and the worship of the Hindu gods and goddesses as shirk, and directed fellow Muslims to stop paying them.[5] It was actions such as this that led the zamindars to consider Shariatullah not only as a simple religious reformer, but also as a political opponent.

Shariatullah's efforts towards religious reform came at a time when the agrarian economy in Bengal was undergoing a remarkable transition. By 1830 the impact of the industrial revolution on textile production in Britain was being fully felt in Bengal. Imported cottons quickly took over from local products, leaving local producers uncompetitive in the market. This, in turn, caused widespread unemployment or reductions in the earnings of weavers that brought their wages below starvation levels. About the same time the impact of the worldwide depression reached Bengal, resulting in the collapse of major agency houses or financiers in Kolkata. This caused a reduction in the flow of credit and drastic falls in prices. Many peasants were coerced into growing an 'unremunerative cash crop in return for paltry advances from indigo factories at a time when rental demands continued to be stringent'.[6] In these circumstances, Shariatullah's call for resistance to the unofficial

Figure 4.1   Portrait of Haji Shariatullah

taxes imposed by the zamindars was enthusiastically greeted by both peasants and weavers. Shariatullah was reported to have command over 12,000 weavers while his 'new creed' bound the Muslim peasantry 'together as one man'.[7] Shariatullah began to be perceived not only as a spiritual saviour; he was also accused by some zamindars of setting up a kingdom like that of Titu Mir.[8]

It was Muhsin al-Din Ahmad, alias Dudu Miyan, a son of Haji Shariatullah, who consolidated the budding Faraizi movement. Dudu Miyan returned from Makkah in 1837 after completing his education and took charge of the movement in 1840 following Haji Shariatullah's death.[9] By 1840, the position of the peasants was changing, the impact of the depression was beginning to subside, the price of commodities was rising and the government's favourable attitude to primary producers of agricultural commodities was becoming apparent. Land reclamation continued extensively beyond the permanently settled areas. Dudu Miyan's world, therefore, was larger than his father's and he found far more followers around him, which encouraged him to radicalize the idea of peasants' rights as well as to take up more action-oriented policies instead of passive mobilization. He proclaimed that '*Langol zar, zami tar*' ['land belongs to him who owns the plough'].[10] He brought before the peasantry the idea of God's sovereignty on earth by proclaiming that the earth belonged to God, and that no one other than the lawful government had the right to rule or impose taxes on it.[11] Thus, under Dudu Miyan the Faraizi creed was transformed into a full-fledged agrarian movement.

Dampier, a commissioner of the Bengal Police, estimated that Dudu Miyan had about 80,000 followers,[12] while Alexander Forbes, editor of the *Bengal Hurkaru*, calculated the number to be around 300,000.[13] On one occasion when Dudu Miyan was on trial in a Faridpur court for his alleged involvement in an attempt to kill an indigo planter, an estimated 3000 boats filled with his followers stood ready at a river nearby to protect him in the event of a guilty verdict. A witness to the Indigo Commission noted that so great was the possibility of unrest that an entire regiment in Dhaka was ordered to stand ready to tackle any incident. It was commonly believed that in this particular case Dudu Miyan's acquittal was the 'result of fear on the part of the Government'.[14] During the Sepoy uprising of 1857, Dudu Miyan was arrested and imprisoned as a pre-emptive measure. The story went that he would have been released if he had not boasted that he could summon 50,000 people at any time to march in whatever direction he so desired.[15] Gastrell, during his survey of three districts of eastern Bengal between 1856 and 1862, noted that the number of the Faraizis was 'annually and steadily increasing'.[16]

The popularity of the Faraizi movement was so great that Dudu Miyan was obliged to divide Eastern Bengal into different areas, called 'circles', and appoint a khalifah, or representative, to each of them. The duty of a khalifah was to enlist new supporters, collect subscriptions for common welfare funds, dispense justice, organize and train armed bodies, set up schools and to teach the proper performance of Islamic rituals. Every ten villages constituted a circle, and the khalifah reported directly to the holder of the next rank in the hierarchy. Khan has examined the Faraizi khalifat system in great detail and has observed that it had a complete and functioning hierarchical structure. At the top of the hierarchy was the ustad or the chief of the Faraizis. Below him was the uparasta khalifah (superior representative); next to him was the superintendent khalifah; and at the bottom was the gaon khalifah (village representative).[17] Dudu Miyan was reported to have received 'applications and memorials like a sovereign from their disciples, fined them and judged and decided their disputes'.[18] This hierarchical network, beginning at the grass-roots level, contributed to the sustainable functioning of the Faraizi movement. (Figure 4.2 shows the extent of the Faraizi movement in undivided Bengal.)

Dudu Miyan was succeeded by his son, Noa Miyan, who maintained the integrity of the Faraizi movement with an influence equal to that of his father. Navinchandra Sen, a contemporary Bengali sub-divisional officer, noted in his memoir that the words of Noa Miyan were like 'Veda' for the Faraizis and that he established 'his own kingdom on that of the English'. Noa Miyan successfully retained the institutional

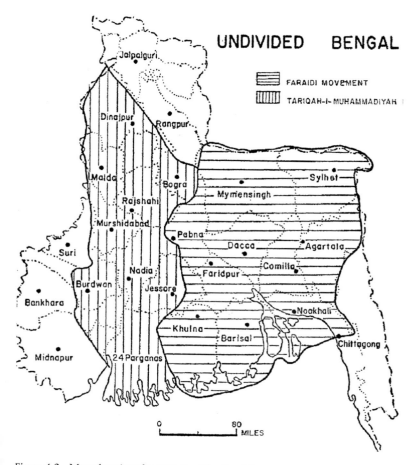

Figure 4.2 Map showing the extent of the Faraizi movement
Source: Muin-ud-Din Ahmad Khan, *History of the Fara'idi movement in Bengal, 1818–1906* (Karachi, 1965).

arrangement of panchayet developed by Dudu Miyan. In case of any disputes in a village, nobody could go to the dewani (civil) or faujdari (criminal) courts. Everybody had to go to the panchayet first; otherwise he was dharmaschuto (excommunicated). Navinchandra Sen noted that the secret of his success as an administrator in the Faridpur district was his alliance with Noa Miyan for mutual assistance.[19]

The number of Faraizi followers continued to grow during Noa Miyan's time. In an essay on Islamic revival, published in the London *Times* in 1873, it was claimed that year after year Islam was converting 'hundreds of thousand of the Indians, especially the natives of Bengal, to the faith of Koran'.[20] In response, Isaac Allen wrote in the *Calcutta Review* the following year that he did not believe that conversion was taking place on such a scale. But Allen thought that a 'possible explanation of the above errors might be found in the rapid conversion of Sunnis and Shiahs into Ferazis or Wahhabis, which had taken place during the last 20 or 30 years'.[21] This picture of the numerical strength of the Faraizis was matched by the claims of their adversaries that each of the three Faraizi leaders had acted as a sovereign or established a 'parallel government'. Some recent researches support similar assumptions.[22] The considerable popularity of the Faraizis raises two important questions. First, how can we explain the emergence of an Islamic reformist-puritan movement under environmental conditions that historians consider conducive to syncretism? Second, how could the Faraizis enjoy such popularity at a time when the political and military power of the British Raj was at its height?

It appears that both the Faraizis and the British in Bengal had their own particular weaknesses and strengths and that these informed their approach to one another. The Faraizis neither regarded the British Raj as a comfortable ally nor stood in outright opposition to it. They made it clear from the beginning of the movement that they did not acknowledge the sovereignty of the British Raj. This attitude was reflected in their proclaiming Bengal as the *Dar-ul-Harb* or zone of war and their declaration that the weekly Friday jummah prayer was not valid under foreign rule. The Faraizis retained this theoretical position until 1947.[23] Yet the Faraizis never openly declared jihad against the British government.[24] This position was reflected in their careful avoidance of any gesture that would identify them with the radical Muslim reformers whom the British government designated as Wahhabis. The Faraizis 'repudiated' the name of the Wahhabis and refused to pray behind a person belonging to this extremist group, or even to eat and drink with them.[25] During their uprising in the North-Western Frontier province, the Wahhabis persuaded some people from East Bengal to donate money and join the battlefield in northern India, but eventually they found that they had been 'cheated', and 'ran away' without ever actually entering the battleground.[26]

The Faraizis' ambivalent attitude towards to the Raj was possibly reflective of their reading of the recent history of Titu Mir's jihad against an increasingly powerful colonial system, which had ended in disaster. There is reason to believe that the Faraizis had sympathy for the cause of Titu Mir,[27] but they never launched armed resistance directly against the Raj. Instead they concentrated on broadening their social base and finding ways to act against any threat to their agrarian domain.[28] Among the Faraizi supporters there were merchants, landholders, landless peasants, government employees, schoolteachers, beggars, vendors, peons, priests, servants, gomasthas – all drawn from Hindu, Muslim and Christian communities.[29] Alexander Forbes testified to the Indigo Commission that religion had nothing to do with the influence of Dudu Miyan and that this could be understood from the fact that one of the Faraizi factions was then led by a Hindu.[30] The inter-community networking was reflected in the non-interference of the Faraizis in the affairs of other religious communities.[31] The movement seemed to have consolidated itself across a strong class base through shared antipathy to all categories of the landed elite in Bengal.[32] Some historians have tended to depict the Faraizi leaders as representing the jotedar or richer segment of the peasant society who mediated between the landed elite and resisting peasants at the lowest level and thus secured political as well as economic benefit from the mediation.[33] However, closer examination of the movement indicates that it remained a populist one and represented a peasant society that was largely undifferentiated. Throughout most of the nineteenth century, the jotedar class itself had no considerable strongholds in deltaic Bengal. It might have been possible for the zamindar to try to bribe the headmen of a village or 'superior' peasants in order to keep a popular agitation under control and to maintain a situation favourable to the government;[34] although in the context of intense peasant solidarity through different 'unions', such manipulative policies would rarely have succeeded. In the Dhaka division, when a zamindar succeeded in bribing some village headmen to swear 'anything against the men or strike', some of them soon found their houses 'burnt down about their ears'.[35] At the same time, potential collaborators in the form of headmen or petty talukdars found it difficult to pursue legal battles and because of financial constraints and 'extreme anxiety', the talukdars compromised with the raiyats.[36] Thus it seems that the Faraizis were able to build a horizontal coalition between peasants of different religious and social backgrounds against any perceived threat from the landed elite or other 'outsiders'. The broad social basis of their political movement put them in a position of relative strength in their bargaining with the government.[37]

The government response in turn was influenced by several factors. It was under compulsion to recognize the efforts of the raiyats to settle on new lands in the khas mahals or chars and forests so that the process of commercialization of agriculture kept pace with the demand for agricultural produce in the market. The government also wanted to ensure a more flexible rent structure, and this could be done through a policy of dealing directly with the peasants at the expense of the zamindars. It is in this particular context that we find the Faraizis flourishing in the chars and forests of the khas mahal areas as well as the chars of the permanently settled areas where the zamindar's authority was weakened or doubtful.[38] Since the Faraizis represented the enterprising peasants in the khas mahals, the avowed government policy of engaging with the peasants rather than the zamindars went in favour of the Faraizis. The cry of the zamindars against the raiyats in general and the Faraizis in particular was, therefore, often received by the government with shrewd indifference. Another factor that restrained the British administration from effectively reining in the Faraizi activities was bureaucratic incapacity. A *Times* report observed that not only was the civil service of Bengal inadequate in performing its duties, but its strength was 'positively diminished'. The government was compelled to farm out the khas mahals to 'the injury both of Government and people' and the manpower was 'scarcely sufficient to collect the ordinary revenue'. There were 230 civil servants attached to greater Bengal, which meant that there were only five officials in every district. The reporter claimed that the responsibility for this situation lay with the people in London who 'in a timid dread of expenditure put off the evil day, try every possible palliative, and postpone the great reform so imperatively required'. He noted that while the number of subdivisions had increased from 25 to 38, the number of uncovenanted officers in the judicial and revenue departments had decreased by two, from 117 in 1836 to 115 in 1855. Within this inadequate pool of officers, interns or less experienced individuals were taken out of consideration, for, the reporter wondered, whether 'anything [would] be more permanently injurious to our rule than the appointment of a young man wholly inexperienced to the management of a district swarming with Ferzees [Faraizis]?'[39]

Though the higher ground of the theoretical debate about the legitimacy of the British government in India remained unresolved and the colonial state's attitude towards the Faraizis remained undefined, the Faraizis managed to engage the Raj in a nuanced relationship of 'mutual interests' largely at the expense of the mediating elite and similar agencies that operated on behalf of the colonial state. But this largely undefined and uncodified relationship between the peasant power and the state has to be understood not in terms of any egalitarian ideology on the part of the administration, but through an appreciation of the ecological regime of the region that built the social and economic bases of the Faraizi movement. The next three sections will examine these ecological dynamics of the Faraizi movement through its relationship with the indigo planters, zamindars and leaseholders of the Sundarbans.

The Faraizis and the indigo planters

The best indigo lands, according to several witnesses to the Indigo Commission, were those that were subject to inundation during the rainy season and that remained submerged for two to three months. The cultivation of indigo was, therefore, mostly confined to the low-lying char lands, for which planters were always on the look-out (Figure 4.3).[40] Indigo was cultivated in these lands in two different sessions. The winter or October session started as soon as monsoon water receded from the char-diaras, and the indigo plants were ready to be cut in May or June, before the char-diaras were once again submerged by the rising water during the next monsoon. The spring sowing started after the first shower of the season, which brought jow (optimum moistness and coolness) to the soil. However, the rising price of rice brought competition for the char lands, inducing raiyats to settle and cultivate these areas.[41] Reflecting the system of cultivation of indigo, aman rice in the delta was sown at the beginning of the rainy season and was reaped in the winter or between November and January. The aush variety was sown after the first shower of spring and was harvested in July or August. Consequently indigo and rice became rival products contending for the same land.[42] It was a general practice of the planters to compel the raiyats to sow indigo after the first shower. They were then made to weed the indigo fields and, when the rivers began to rise, they were forced to cut indigo plants.[43] This put the raiyats into a situation where they lost their best lands and missed the opportunity to utilize the right season to cultivate ecologically appropriate and commercially profitable crops on whatever lands were left over for their own use. One raiyat testified to the commission that of the ten bighas of his land six were taken for indigo, and the remaining four were allowed

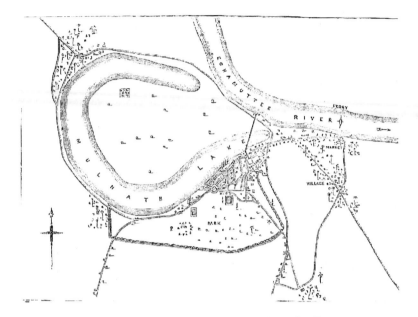

Figure 4.3   Sketch of river and char-centric surrounding of indigo concern
Source: Colesworthey Grant, *Rural Life in Bengal* (London, 1860).

for cultivation of paddy, which yielded nothing as the sowing was 'unseasonable'.[44]

Gastrell, a contemporary surveyor of land, remarked that no people were so wedded to dastur (old customs) as the peasantry of Bengal. When all peasants looked forward to sowing their rice with the first proper shower of the spring, the planters also required help of the peasants for sowing indigo. Thus the interests of the raiyat and the planter clashed. Since the planter had often forced the raiyat to put his labour at the service of the cultivation of indigo, the dastur which the raiyat loved was sacrificed to indigo, which he hated.[45] Gastrell also noted that if the rice crop failed where the raiyat had had nothing to do with indigo, he blamed kismet (fate) for his loss. But if he cultivated indigo, most of the blame shifted from kismet to indigo. This was because the raiyat 'had perhaps to use his ploughs at a time, when, otherwise, he would have used them solely for his rice lands'. Argument in such cases, Gastrell observed, was simply useless: indigo interfered with rice and the raiyat did not like it for 'this fact alone'.[46]

New chars were overgrown with doob (tender grass) that provided plenty of food for cattle, but as these lands were increasingly appropriated by the planters, so access to this pasturage was denied to the tenants and the goalas. Frustrated Bengal peasants and goalas, who held their cattle as means of livelihood but also in great humane affection,[47] would deliberately send their cattle to graze in the doob that grew with the indigo, mostly at night, with the animals protected by armed bands.[48] The planters would retaliate by imposing taxes based on the prints of cattle hoofs on the plain. If a single cow strayed into the indigo field, an exorbitant fine was imposed on the whole village from which the cow came.[49] Dissatisfaction among the raiyat was further deepened by the impoverishment of the land caused by indigo cultivation. One witness to the commission revealed that the yield in certain rice fields had fallen by two-thirds.[50] Thus the cultivation of indigo deprived the raiyat of the two main ecological endowments that provided subsistence and commercial production in deltaic Bengal: the best land and use of the right season.

The Faraizis responded to these challenges with strategies ranging from the formation of peasant unions to outright violence as well as through the use of religious signs and symbols to mobilize different communities. Given the widespread discontent among the goalas, who were mostly Hindus, the Faraizis sought alliances with them,[51] and the goalas resorted to Islamic signs and symbols in the development of a collective strategy for resistance. A planter observed that he knew an old Muslim raiyat who had stopped indigo cultivation. On asking why he did so, he replied that the villagers had formed a coalition against the planters through kissing the Quran and that one village could not settle without the other.[52] Such networking looked formidable in the Barasat subdivision on the specific question of the measurement of land. In the Charghat Pargana of Barasat in central Bengal, a fertile diara-char region, Faraizi resistance was not directed against the actual measurement of land selected for indigo cultivation, but against the measuring rope by which the planter attempted to acquire more land per unit of measurement than the traditional measurement allowed. Ashley Eden described the Barasat raiyats as the 'most intelligent set of ryot' that he had ever met and noted that the Barasat raiyats being Faraizi had a 'complete organization'. They used frequently to meet together to exchange ideas and were in constant communication with the merchants of Kolkata; they knew all that was going on in the neighbourhood.[53] Following his visit to Charghat, Eden reported that he was fully convinced that if such measurement persisted, there would 'be one of the most serious affrays that ever took place in Bengal'. He further noted that Charghat contained several thousand Faraizis who all 'banded together to prevent any interference with their rights, real or supposed'.[54] During the widespread resistance against indigo cultivation in the late 1850s, the Faraizis used the weapon of hookah-pani-bandh or social boycott to strengthen the coalition of peasant interests.[55] Blair B. Kling has remarked that the Faraizi peasants who participated in the indigo disturbances were a 'tightly organized puritanical sect' and were skilled in 'military organization and the use of arms'.[56] The Faraizis, along with other indigo cultivators, were able to threaten the indigo interests and the government was forced to scale down the power of the indigo planters and the cultivation of indigo in Bengal.

One of the frequently cited incidents directly related to the Faraizis was the killing of the gomastha of an indigo establishment, managed by a European by the name of Dunlop. According to Faraizi sources, this gomastha posed as a 'little zamindar', forcing the raiyats to plant indigo on their best rice-lands for only a nominal remuneration and punishing those who refused to comply with the planters' conditions with chili powder and a beard tax. About the same time, in a raid launched at the instigation of Dunlop and the said gomastha, property was seized from the house of Dudu Miyan himself and four of his associates were killed. This incident was followed by the murder of the gomastha.[57] Contemporary writers tended to explain the incident as an act of personal revenge on the part of Dudu Miyan.[58] However, the testimony of Musli Hawaladar, one of the Faraizi prisoners arrested along with Dudu Miyan, told a different story. He informed the court that char Kamraj, where he lived, had formerly been leased to Dunlop. When his term expired, Dunlop wished to renew the lease, but Musli and certain other cultivators objected, petitioning the collector of the district and asking that the lease to be given to another interested European, Henry Oram. Musli claimed that Dunlop had been looking to revenge himself against those who had petitioned the government in this regard.[59] This testimony revealed that the Faraizis worked determinedly against any challenge to the settling of khas mahal lands in their favour. They did this not only through the use of force, but also by exploiting opportunities that arose from rivalry between the 'outsiders', such as the indigo planters.

Kling remarks that even if there had been no indigo-related disturbances, the industry would still have died out in lower Bengal. He saw both the gradual increase of the land level, which resulted from the eastward shift of the Ganges, and the decline in the demand for indigo in the world market as contributing to such decline.[60] This argument needs to be treated with caution. In response to ecological changes, the indigo plantations could have been shifted to the more active part of the delta in the direction of the shifting river and even beyond, to where the rivers Padma, Brahmaputra and Meghna jointly watered the landscape, creating land that was consistently suitable for indigo cultivation. But indigo did not travel in that direction. As far as demand for indigo is concerned, it continued to be planted for quite a long time in Bihar and other areas after its cultivation fell substantially in Bengal after 1860–61. Production and export statistics do not show a decline in the demand for indigo in the world market until at least the following decade.[61] This has prompted David Hardiman to ponder why neither Kling nor Ranajit Guha, two major historians of the anti-indigo resistance, focused on the question of why indigo continued to be planted in Bihar and other areas. Hardiman sought to explain this in terms of the distance of Bihar from Calcutta, a place which was a 'relatively remote and backward region in which such oppression was less easily observed by what passed for "public opinion"'.[62] But it appears likely that indigo cultivation could not extend to the more active deltaic eastern Bengal in part because of the greater number of the Faraizis in these areas where rice and jute continued to be the major commercial agricultural products. Following their successful resistance to indigo interests in central Bengal, the Faraizis were emboldened to stand against any effort to extend the cultivation of the plant. Robert Morrell, for instance, testified to the Indigo Commission that in one of his packets of seeds sent from Kolkata there were indigo seeds which were sown by his gardener 'by mistake'. He was in Kolkata at that time, and when he returned, the raiyats 'insisted on having it [indigo plant] pulled up and thrown away'. Morrell complied and destroyed the plants.[63]

The Faraizis and the zamindars

Conflicts between the Faraizi peasants and the zamindars mostly involved the chars and diara lands and surfaced in two distinct ways. First, as we have seen in previous chapters, the destruction of older tracts of lands by the diluvial action of the rivers that skirted their estates entailed great loss to the zamindars that was not compensated for in the form of new land created by alluvial accretions. Instead, these lands legally became the property of the state.[64] They also became the 'favourite retreats' of the Faraizi raiyats.[65] The zamindars argued that they had rights over the khas mahals that were created off their own estates and attempted to prevent the Faraizis from settling there, often with the help of professional lathials or clubmen. The Faraizis responded with equal violence. Second, newly emerged chars often became the subject of ownership battles between two or more neighbouring zamindars, resulting in lengthy legal procedures. While the arguments continued, the Faraizis settled on the new lands and often paid no rent at all until the lengthy legal procedures were completed. When the law suits were decided the Faraizis were still 'not very willing to pay much' and banded together to 'resist every demand for any increase however reasonable it may be'.[66] The raiyats who had settled on a char not only asserted their ownership, but claimed to have permanent occupancy rights on it. The introduction of Regulation X of 1859, which conferred right of occupancy on a raiyat if he had possessed a particular piece of land for more than 12 years, consolidated the raiyats' dual claims of possession and ownership rights. This class of occupancy raiyats, as Richard Temple observed in the 1870s, 'were constantly growing in numbers and representing a larger and larger section of the ryots'. On the basis of his occupancy right, the raiyat tended not 'to pay more than the old established rates of their part of the country'. The relative absence of agrarian unrest during the 1860s was probably a result of the introduction of the regulations of 1859 which consolidated the Faraizi position in those lands they had reclaimed and settled.

Peasant unrest, however, had a remarkable resurgence at the beginning of the 1870s. Leagues and unions were formed sporadically in East Bengal and these continued to be active until 1885 when the Bengal Tenancy Act was passed.[67] Partly because the battle over indigo was virtually over and partly because the acts of 1859 were largely in their favour, between the 1859 acts and the resurgence of unrest in 1870, the Faraizis generally opted for legal options against the landlords.[68] The 'first feeling' of the raiyats, as a government secretary wrote, was 'almost always' a sort of rebellion against the system (the Permanent Settlement) that had limited the claims of the government on the zamindars, but had 'left without limit the claims of the zamindars on them'. The raiyats asked why they should go on paying continually increasing rents to the zamindar when the zamindars themselves paid no more to the government. Consequently, whenever there was any degree of Faraizi or other religious revival 'among the pugnacious Mahomedans of the eastern districts' it always took 'the shape of a struggle about rents'. The government observed that the zamindars, being aware of their own lack of power to turn things in their favour, stopped resisting the growth of rights of occupancy. They generally compromised on a moderate increase of rent or offered leases at 'rents fixed for ever'. Thus, according to the government, if eastern Bengal was saved from an eventual serious crisis, it was through the gradual operation of a process by which a great number of petty proprietors were created and the security of the state was enhanced on the 'broad base of a widely extended distribution of property'. In this connection the Lieutenant-Governor of Bengal, George Campbell, thought that if the government of India had made a 'hideous mistake' in resisting the accrual of occupancy rights, then 'quiet Bengal must soon be held by a great army, 'if it is held at all'.[69]

The government of Bengal seemed to have encouraged the Faraizi movement against the zamindars who wanted to prevent the accrual and enjoyment of occupancy rights. In their turn, the Faraizis consolidated their position by voluntarily paying the actual amount of rent due to the government. The Administrative Report of Bengal of 1871–72 mentioned that in East Bengal whenever a landlord had quarrels with his tenants, it was a fashion to 'stigmatize the latter as "Farazees", a sect professing reformed tenets, and doctrines of equality, and to attribute to their conduct a political character'. Such was the case when, in two separate incidents, one in Barisal and the other in Dhaka, disputes occurred between the tenants and the European lessee of a large government estate. But the government found that the demand of the landlords for an enhanced rent was not legal. The Lieutenant-Governor, therefore, dismissed the necessity of taking extraordinary measures against the Faraizis in this connection. The result was that they 'paid quietly old rents which they were legally bound to pay' and they further said that they were ready to contest the question in the courts regarding the future.[70]

The peasant movement in Pabna in 1873, however, turned remarkably violent. The outbreak was caused by five zamindar families who had planned to undermine the occupancy rights of the raiyats. One landlord tried to put an end to all opposition by forcing his tenants to sign and register agreements to give up their rights of occupancy.[71] Pabna resistance also coincided with an economic recession followed by a sudden fall in the demand for jute in the world market. Pabna (Sirajganj), as one of the largest jute marts in the delta, was severely affected by the depression. The Faraizis were both producers and traders in jute[72] and the landlords' demand for illegal taxes and resistance to accrual or suppression of occupancy rights during this time of depression worsened the situation. More than 30,000 people banded together and both the Faraizis and Hindu raiyats waged resistance, as R.C. Dutt observed, together as 'ryot and rebel'.[73]

Through the Pabna movement the raiyats were able to obtain considerable concessions from the landlords. One zamindar let the land for ever at fixed rates and a number of zamindars promised that they would comply with whatever terms the unionists liked.[74] The zamindar whose attempts to force the raiyats to give up their occupancy rights proved to be one of the immediate causes of the Pabna resistance was said to have lost three lakhs of rupees in the legal battle with the raiyats.[75] The Pabna movement was quickly followed by similar movements in other parts of eastern Bengal where zamindars tried to curtail the raiyats' rights on land. A land revenue report of 1874–75 noted that a 'spirit of continual resistance' by the raiyats had manifested itself in several places, mostly in the char areas. During his visit to eastern Bengal in 1875, Richard Temple was impressed by the overall social and economic condition of the region. However, he expressed his concern that the landlords' attempts to collect rent by force, to increase rent and to impose certain unauthorized taxes could soon lead to intense political turmoil in the districts of Dhaka, Faridpur, Tippera and Barisal.[76] Temple's apprehension was justified. Peasant unrest continued to spread, not only in those districts, but also in other districts of the delta, such as Mymensingh. On many occasions, Noa Miyan himself led the movement.[77] Given the strong and broad base of the movement against 'high landlordism' of the late nineteenth century, the government felt it necessary to enact the Bengal Tenancy Act of 1885, which reasserted the provisions of peasants' rights of occupancy on land as set out in Regulation X of 1859.

The Faraizis and the leaseholders of the Sundarbans: a case study

Despite the extensive land reclamation and settlement process that took place in the Sundarbans throughout the nineteenth century, peasant unrest in the Sundarbans did not appear to be as intense as it was on char-diaras further inland.[78] The relatively low intensity of agitation in the Sundarbans may be attributed to the fact that the large leaseholders who were entrusted with reclamation of forests and settlement of peasants on behalf of the government did not interfere with the rights of occupancy accruing to the cultivator or with his selection of crops. A leaseholder was focused on ensuring profit against his investment in reclamation. The main conflict between the leaseholder and the cultivators, therefore, was centred on the question of periodical increases in rent. As long as the leaseholder did not demand too great an increase in rent, there was no trouble, but peasant action followed promptly when he overstepped the line. What follows is a case study of the changing relationship between a reclamation company owned by the Morrell family, the largest leaseholder of the Sundarbans, and its raiyats, who happened to be mostly Faraizis.

Robert Morrell, the principal shareholder of the Morrell Company, took lease of a number of large plots in the Sundarbans around 1850, an area which gradually became known as Morrellganj. From the start of the lease, Morrell maintained a policy of offering a variety of material incentives to encourage the raiyats to reclaim and cultivate forest areas. He offered loans to the raiyats but did not sue them for arrears and in some years he remitted loans for sums as large as Rs. 5000. He used to arrange 'feasts' for the raiyats.[79] The measures that he took to prevent his staff exacting undue influence over the raiyats also played a decisive role in bringing relative peace to Morrellganj. Morrell initially started paying his staff, which included his administrative, revenue-collecting and security personnel, salaries ranging from Rs. 4 to 12 per month. When it was found that members of his staff were coercing the raiyats to pay more than the rent fixed by the Morrells, the raiyats petitioned Morrell, offering an additional two annas a bigha if he stopped his staff from taking nazr. Morrell understood that the raiyats' offer to pay money over and above the original rent came from their appreciation that the behaviour of the staff was provoked by their own poor salaries and that it could be stopped if their salaries were raised. The offered arrangement was seen as mutually beneficial. Morrell raised his staff's salaries three or fourfold, there were no longer complaints of extortion and the raiyats paid their rent regularly.[80] This incident exemplifies both the financial acumen of the peasants and their understanding of the well-being of fellow members of society, even those representing the state mechanism. Morrell seemed to have appreciated the social logic of reciprocity and his style of managing some Sundarbans khas mahals was considered so successful, particularly in dealing with the Faraizis, that he was offered the status of an honorary deputy magistrate and authorized to supervise the collection of rent of some neighbouring estates as well.

The cordial relationship between the Morrells and the Faraizis was exemplified by an incident in the 1850s. The Sundarbans estate of Tushkhali consisted of 103,613 bighas of land and contained more than 10,000 raiyats. At the expiry of the existing lease of Tushkhali in 1859, J.H. Reily, the commissioner of the Sundarbans attempted to lease the estate at a greatly increased rent of Rs. 119,743, up from Rs. 39,149. The increase enraged the raiyats, who joined together to protest against it and – or so it was reported – even wanted to shoot Reily.[81] The situation became so volatile that Reily called for the assistance of Robert Morrell, who pacified the raiyats by his 'goodwill'. At the same time, Reily indicated that the new rate of rent was fixed and final and that no more increases would follow.[82]

When the Board of Revenue continued to the next round of leasing the Tushkhali estate to the highest bidder, Robert Morrell came forward to demand preferential treatment in obtaining the lease. He argued that when he took charge of the estate as an honorary deputy magistrate, the entire estate was in turmoil and he had brought stability there. Morrell insisted that he ensured the collection of rent in its entirety and thus enhanced the value of the property which, owing to the 'riotous character of the tenants', was otherwise regarded as a risky financial investment. He cited the example of Rajah Sutchurn Ghosal, who refused to compete for the lease of the estate on this account. Morrell also argued that his involvement with this estate had placed him in such a position in the Sundarbans that he could no longer withdraw from it without seriously affecting his interests in the other Sundarbans estates that belonged to him. Morrell reminded the Board of Revenue that a leaseholder owed much of his success to the roab or goodwill he acquired 'amongst an ignorant population'. He feared that having once included these estates within the sphere of his influence, a retreat from them or renouncing them to any other person would seriously affect his interests. Morrell also noted that he responded to the government's call for assistance to tackle the 'riotous conduct' of the raiyats and ensured the collection of rent on the understanding that if he succeeded in collecting the rent and managing the estate satisfactorily he should thereafter be entitled to a preference in any future leasing of the estate. Morrell lamented that but for this understanding he would never have consented to devote so much of his time to the management of these estates on the utterly inadequate remuneration of 10 per cent, which had proved insufficient to maintain his establishment and which had left him nothing for his trouble.[83]

While Morrell petitioned the government as a successful collector of rent and manager of some khas mahals, his success, in fact, was based on his understanding and agreement with the raiyats that Reily's enhanced settlement rate of 1859 was indeed 'fixed and permanent'. Morrell, therefore, also reminded the board that any rent increase beyond the ceiling fixed by Reily – a considerable likelihood if the lease was given to anyone other than himself – would be 'a breach of faith' with tenants of the estate.[84] Reily himself recommended Morrell's application for the lease of Tushkhali by reminding the Board of Revenue how Morrell stood by him when he was under pressure from the Faraizi raiyats.[85] In response, a secretary to the government of Bengal found the ground on which Reily pressed Morrell's claim as 'doubtless of great weight'. The administration noted that it had not overlooked Morrell's services and in recognition of this, he was sanctioned a lease for ten years and the payment of an additional 10 per cent of the collections.[86] During the following ten years of the lease, the Morrells never attempted to raise the rent, by recourse to law or otherwise.[87] In consequence, no remarkable peasant unrest was reported during the period of this lease.

At the next renewal period of the Tushkhali lease, there were several influential parties interested in making bids. Among the contenders was the Nawab of Dhaka, Khaja Abdul Ghani, who offered to pay Rs. 200,000, or more if required, which was significantly more than the amount offered by the Morrells. A government secretary asked whether the Morrells had done anything to deserve the continuation of the lease and whether the government should sacrifice a huge amount of revenue by giving the lease to the Morrells instead of to Nawab Abdul Ghani.[88] On the other hand, support for the Morrells was also strong. One of the five reasons put forward to support the case for settlement with the Morrells was that they had shown 'peculiar tact' in dealing with the 'unmanageable' Faraizis.[89] At the same time it was reiterated that the Faraizis considered Reily's 1859 settlement as fixed and permanent.[90] The lobby within the government in favour of the Morrells' case therefore cautioned against settling Tushkhali with Khaja Abdul Ghani even at a jumma 'as high as it might be', because such an 'unfettered purchaser' would only bid high with a view to increasing the raiyats' rent.[91] In the end, Morell's 'kindness', 'spirit of improvement' and 'tact' won over the highest rental bid offered by the Nawab. The Tushkhali estate was leased to the Morrells in 1871 for twenty years.

Yet the Faraizis soon began to mobilize violently in response to a series of new developments that put the government, the Morrells and the Faraizis in a different situation. While in the 1871 settlement the previous rental was left unchanged, the government had agreed the lease only on condition of a security deposit of Rs. 15,000. This was based on the understanding that on the completion of the settlement, the Morrells would be allowed to raise the additional amount.[92] The Morrells, now represented by two younger brothers of Robert Morrell, who had passed away in the meantime, thought that they could collect the extra amount through rent increases and, hence, agreed to the government's conditions. But within a few months of the settlement, the raiyats of Tushkhali had united against the proposal of increased rents. The raiyats sent a deputation to Kolkata to ascertain what rights they possessed and therefrom learnt that the rent could not legally be increased.[93] The raiyats then decided not to pay any additional rent. The Morrell brothers asked the government to intervene and help collect the additional amount, but the government did not think it advisable to intervene. The collector of Barisal, Henry Beveridge, observed that the Morrells had bargained with the government to pay an increase of Rs. 15,000 over the old rental because they assumed that they could easily collect this sum from the raiyats; but they were mistaken, and in consequence were losers by the estate. There was no reason, Beveridge thought, why the government should interfere and attempt to resettle the issue which the leaseholders had failed to accomplish. Beveridge advised that the Morrells give up the lease.[94]

In their response to the strong resistance of the raiyats to rent increases and the government's suggestion that they give up the lease, the Morrells wrote to the government that they had taken the farm at a higher jumma of Rs. 15,000 in the understanding that raising the necessary sum would not require coercion, neither had they the power to coerce the raiyats. The Morrells believed that if they had succeeded in raising the rents, they could have gained some profit and a collateral advantage in that they could secure labour at fair rates for the reclamation of their still unreclaimed forest lands. But, the Morrells noted, they had found it impossible to convince the raiyats to pay even the lowest increase, because they were 'under the erroneous belief' that the dowls (settlement) granted to them by the government (Reily) could not be altered by a leaseholder. Thus, in a radical reinterpretation of Robert Morrell's reading of Reily's 1859 settlement, by which the raiyats had stopped their agitations, the Morrells informed the government that they did not believe that Reily's settlement was in fact permanent. The Morrells observed that if this assumption was doubtful – and both the Board of Revenue and the commissioner of Dhaka had 'serious doubt on this point' – and consequently that the 'impossibility of enhancement by a farmer was certain', then the settlement should not have been made with them 'without first settling this point with the ryots'. Finally, the Morrells proposed to give up the estate and requested to relieve themselves of the liability of paying the government the previously agreed sum of Rs. 15,000.[95] The Bengal government immediately accepted the proposal and brought the estate under khas management. On the question of a fresh settlement of the estate, the Lieutenant-Governor, Richard Temple, suggested that the revision should allow the just share of the government as landlord in the increased profit of cultivation, but the demand should be 'scrupulously and unquestionably moderate'. A demand thus fixed, Temple continued, should be exacted in a 'quiet and considerate' manner. Temple noted that their consideration did not reflect the government's weakness and it would be 'so firm indeed' that the raiyats who were of a 'somewhat turbulent character' might see that the government was not going to be trifled with. Finally, however, Temple emphasized that the demands the government meant to enforce should be so 'mercifully moderate as to be beyond question as to equity'.[96]

Decline of the Faraizi movement and beyond

After the death of Noa Miyan in 1884 the Faraizi movement appears to have abandoned its original agrarian agenda and lost much of its political influence.[97] Its peasant-political authority declined further when one faction joined the new Muslim League in the wake of the partition of Bengal in 1905. Considering their remarkable success in agrarian politics for more than half a century, the decline of the Faraizis invites critical examination. Their decline is also paradoxical in the context of both deteriorating economic conditions around the turn of the twentieth century and the rise of rack-renting, which together were able to fuel an even stronger spell of agrarian unrest. Indeed there was peasant resistance throughout the first half of the twentieth century, but its nature was different to that of the Faraizi movement, a faction of which now allowed itself to be subsumed by the communal political programme of the Muslim League.

At first glance, the decline of the Faraizi movement could be linked to the decline of the ecological regime of the Bengal Delta. By the turn of the twentieth century the waterways of the region had deteriorated as the result of a number of man-made and natural causes and the process of deterioration continued throughout the twentieth century – topics that will be examined in Chapters 6 and 7. The decline in the water regime had far-reaching consequences for the Faraizi movement. First, increased waterlogging or abnormal flooding resulted in the decline of the production pattern upon which the agrarian economy and Faraizi mobility depended. Second, the Faraizi communication network was largely built along the waterways of the region. The Faraizi headquarters were so closely connected with different rivers that the Faraizis were able to move across the whole of eastern Bengal with ease and efficiency.[98] The death or fluvial weakness of the rivers caused problems of navigation which could have resulted in the displacement of the indigenous network of information gathering and political mobilization. Third, by the turn of the twentieth century, by dint of superior methods of collecting information on interior landscapes and property, the government seemed to have obtained a relatively clear knowledge of the fluid and peripheral ecological regime in which the Faraizis flourished. Since the state could now more confidently ensure its presence in the interior of the delta, the benefit of peripheral wilderness, in which the Faraizis operated, ceased to exist.[99] The foremost challenge to the Faraizis, however, came not from ecological decline or the state's better knowledge of the countryside, nor from the Arms Act of 1878 which was designed to disarm the Faraizis and the Santals of West Bengal.[100] If the Faraizi movement can be considered a political pressure group that operated through the social and economic dynamics of the nineteenth-century agrarian world, then its decline has to be examined in the context of the emerging domain of nationalist politics, as opposed to 'agrarian' politics, within Muslim society itself. It is also important to examine the new wave of modernization processes – the so-called Muslim Bengal renaissance – that began in the late nineteenth century, the ways in which this positioned agro-Islamic polities within its influential world view, and the social, economic and political consequences of this.

A soft hierarchy had long existed within the agrarian Muslim community, as reflected in the co-existence of poorer and wealthier peasants, such as talukdar or hawaladar. In most cases these relatively superior peasants did not operate as 'external' actors in agrarian communities. They offered leadership and protection to the raiyats in the fluid char landscape or the wilderness of the Sundarbans. They worked in the fields along with the ordinary raiyats and thus remained an integral part of mainstream peasant society. Towards the end of the nineteenth century, however, some of these superior raiyats began to convert their occupancy rights into tenure-holdings. As a tenure holder the raiyat became a 'landlord-raiyat' who would not himself work in the field but would employ under-raiyats on his behalf.[101] The process of empowerment of a section of rich peasants eventually led to the formation of a horizontal alliance between the emerging Muslim bhadralok middle class operating in the cities and the mofussils and richer peasant in the countryside. The growing distance between the bulk of the peasantry and the neo-bhadraloks was aptly illustrated by the following comments by a little-known writer:

By Bangali, I mean the cultivators. Those who are living in Bangla Desh and have done some kind of higher education cease to be

Bangali: some of them become *shaheb*, some become mister, some of them become *bahadur* and so forth. In one way or the other, they have redeemed themselves of the name of Bangali; and to avoid the air of this bad name they have left the village to live in the towns and thereby have totally removed their presence from the Bangali class. They are not Bangali anymore. If I want to mean anything by Bangali I mean the cultivators. These cultivators have no other means to remove the stigma of Bangali from themselves.[102]

As the urban political space of the Muslim 'middle class' expanded, forms of agrarian politics lost ground. This subtle shift took place within an emerging hegemonic order in which the peasants were persuaded by their 'modern' fellow co-religionists that salvation rested in a communal alliance that would prevent the intrusion of the Hindu 'outsider' into their socio-economic space. In this sense, communal consciousness among the Bengal Muslims arrived on the wings of modernity. Understandably such communal mobilization of ordinary Muslim peasants was informed by an urge to neutralize the rural bases of politics in order to widen the urban political space to face the challenge of modernity in general and of the Hindu bhadralok's political ascendancy in particular. The peasant was doubly exposed by this reorientation of the political order. First, as a politically charged 'communal bond' was forged among the Muslim peasantry, both urbanite ashrafs and the rich peasants were able to strengthen their hegemonic grip on the ordinary peasantry. In the context of an apparent neutralization of the notion of 'class-conflict', it was not surprising that the Faraizis, who successfully resisted the Muslim zamindars and oppressive 'rich' peasants in the nineteenth century, gradually became peripheral to the emerging urban middle-class body politic.[103] (Although there are still those who cling to the Faraizi ideology: see Figure 4.4.)

The decline of the Faraizis also coincided with the loss of the social and economic autarky in agrarian East Bengal – a scenario that demands broader understanding of the emerging social power of the 'modern' in which the 'agrarian' was imagined. In the bulk of recent literature, the 'modernist' cultural and political discourses of Muslim Bengal in the post-1857 era, flourishing in the works and activism of individuals like Syed Ameer Ali, Nawab Abdul Latif and Nawab Khawja Salimullah, are considered separately from agrarian history. Yet any meaningful understanding of the decline of agrarian society in East Bengal must depend on identifying the relationship between the 'modern' elite middle class and the 'traditional' rural peasant community. This can be done, as Sumit

Figure 4.4   Photo of a recent Faraizi follower
Source: Taken in 2003 by the author.

Sarkar has shown, by looking at the way in which the urban middle-class Muslim pressure group – we might call them Muslim bhadralok – actually began to represent itself as the 'peasant' in narrating agrarian problems, thus both creating and appropriating the category of the 'nation' in the name of the peasant. Arguments by Partha Chatterjee, Taj Hashmi and Sugata Bose are more explicit in that the nexus of the Muslim elite, middle-class and rich peasantry appropriated the vast majority of the political and productive autonomy of peasants within Muslim nationalist politics. These interventions are important, but by focusing on the emerging communal convergence, they ignore the equally remarkable disjuncture that modernity created between the nationalism of the urban middle class and the worldview of the peasant.

The emergence of the Muslim bhadralok pressure group was reflected in a Bengali tract written in 1889 by Sheikh Abdus Sobhan. Sobhan, whose book was patronized by a number of Muslim barristers, zamindars and munshis, was apparently annoyed by the National Congress's bid to accord parliamentary representation to the Hindus and the half-hearted way it attempted to represent the Muslims. In this context, according to Sobhan, the 'duty of the Muslims' should to be to inform the Raj and 'mother Victoria' about the 'deteriorating conditions' of the Muslims. By advocating the importance of establishing organizational units in the countryside, connecting them to the capital city, raising subscription from all walks of life and publishing English, Urdu and Bangla newspapers, Sobhan's proposals foreshadowed the emergence of the Muslim League. Sobhan lamented the predominance of Hindus and the absence of Muslims in the modern institutional spaces of courts of law, police stations, railways, telegraph and postal services, hospitals and so on. Such narratives of the 'decline' of Muslim society in opposition to the 'prosperity' of the English-educated Hindu bhadralok persisted in Muslim public domains till the partition of India, and is a perception that continued to influence history writings in postcolonial times.[104] What remain unfocused in the discourse of Muslim 'regeneration' and modernization was the way in which it simultaneously developed a sharp criticism of the agrarian domain of East Bengal. For instance, Sobhan's narrative of the 'decline of the Muslims', derived as it was from an appreciation of the 'better' material conditions of the Hindu other, was also notable for its positioning of the 'traditional' agro-religious order in the author's scheme of the modern. He notes:

...these uneducated *aborjona* (garbage), having sold us to the Hindus are roaming around with their *talbe elem* (disciples) around the house of the ignorant peasants for appropriating their cash and crops through advocating for more 'prayer' and less 'eating'... but if they are asked any question about religion they are dumbfounded ... they are greatly harming the poor peasants.

Sobhan laid claim to newer religious knowledge than that of the Muslim agrarian leadership who formed an integral part of the rural polity and agrarian relations. Sobhan argued that strenuous examination had found that those Muslims who had returned from London after studying there were well-versed in the sacred religion of Islam. 'Only a few ignorant, sunk in darkness of orthodoxy, sometimes are calling the former "Kafir" because of their learning of English.' Sobhan asks what they would say of the Turkish and other Muslims of Europe. And he advised his readers to keep away from those who considered the English-educated as Kafir.[105]

This emerging Muslim discourse of modernity shows how the subjectivity of the 'agrarian' along with its cultural, social and economic autarky became lost in a communal-nationalist space defined by an overwhelming optimism about modernity that sought solutions to many of the problems of the 'nation'. Yet, it would be a misreading of history to conclude that this new agency of modern Muslim politicians' hollow, self-seeking Islamization of nationalist politics solely defined the fate of the 'agrarian' in East Bengal. If for the Muslim bhadralok the agrarian was made peripheral to a form of nationalism that aimed for spaces in the techno-bureaucratic and institutional realm of the colonial state, the nationalism of the Hindu bhadralok had a larger stake in putting the agrarian space into the heart of the nationalist project. While for the Muslim elite/middle-class pressure groups, national ambition centred on securing spaces in the modern institutions of education, law, governance and parliament, for their Hindu counterparts, the agenda of returning to the countryside was crucial and in fact 'tradition' came to be an integral part of the modern that defined their forms of nationalism, as recent research suggests. In the next chapter we will deal with the dynamics of the Hindu bhadralok's engagement with the agrarian world after the turn of the twentieth century.

5
Return of the Bhadralok: the Agrarian Environment and the Nation

Historians of modern Bengal have tended to settle the bhadralok squarely in the urban milieu, implying a gradual decline of their rural connections in the late colonial period. This attenuated position of the bhadralok in the historiography of agrarian Bengal – East Bengal in particular – is qualified by the discourse of the 'rich peasant'. These better-off peasants are considered to have formed the dominant group in the countryside by virtue of a peasant 'autonomy', which was too strong for the bhadralok to break.[1] It is suggested that this trend strengthened following the mobilization of the bhadralok into anti-colonial resistance in the mofussals, particularly after the first partition of Bengal in 1905. Some of those who were supposed to have remained in the countryside also made their way to the towns following the depression of the 1930s, others followed in 1947.[2]

This chapter – in contrast – argues that the bhadralok remained an active force in the agrarian relations of eastern Bengal, particularly after the turn of the twentieth century. But the role of the bhadralok cannot be fully appreciated if they are considered merely as a detached rentier group whose role in agrarian changes was limited to increasing opportunities for extracting rents from the peasantry. A more important issue – and one that remains largely unexplored – is precisely the way in which the bhadralok broke the so-called 'autonomy' of the small-holding peasantry. This chapter focuses on the specific agro-ecological regime of the Bengal Delta, which informed the bhadralok's increasing engagement with the agrarian world. The tendency I note here of the bhadralok's return to the rural milieu was different from the tendency of the immediate post-Permanent Settlement generation of the wealthy zamindar-bhadralok who bought up large estates and then largely remained absentee landlords. The new generation of bhadralok, increased in number and diminished in prosperity, could afford neither to buy up large zamindari nor to remain indifferent to and aloof from the primary production process. Below we examine the process and dynamics of the agrarian engagement of the twentieth-century bhadralok in the tumultuous decades of anti-colonial nationalist imagination and politics.

The nation and the countryside: East Bengal and the demography of the bhadralok

It has recently been suggested that in its early phase Bengali nationalism did not have any particular 'programme' and was mainly concerned with 'lamenting for the lost glory of the nation, bemoaning its present plight and soliciting British help'. It was around the turn of the century, especially in the context of the partition of Bengal in 1905, that some distinct nationalist programmes were formulated, including the determination to achieve economic self-sufficiency, political prowess (in the sense of Bankim's bahubal) and self-prowess (in the sense of Rabindranath Tagore's atmashakti).[3] The Bengali terms 'swadeshi' and 'swaraj' powerfully expressed the synthesis of these late-colonial Bengali nationalist programmes. One key feature of the Swadeshi Movement was a firm commitment to return to the countryside. This reflected an attempt to break from the Calcutta-centric urbanization process that coincided with the consolidation of British power.[4] The new form of nationalism was different from the late nineteenth-century patriotic awareness of the countryside in that it was a committed programme that aimed to be implemented rather than merely imagined. The renewed call for a return to the countryside reflected the intellectual and political aspirations of a nation with the countryside at its heart. As Aurobindo Ghose, speaking in the eastern Bengal town of Kishoregunj in 1908, asserted, 'Swaraj begins from the village.'[5]

To situate the countryside and agriculture within the nationalist project required the invocation of an agrarian past that was simply prosperous, an imaginative reworking that Prathama Banerjee called a 'manifesto for practice'. In other words, the evocation of a prosperous agrarian past was not to aspire for or justify historical or chronological accuracy, but to provide a template for the future.[6] In this context, the specific metaphor of 'Sonar Bangla' (Golden Bengal) which flooded Swadeshi literature is significant because the term, most powerfully used in the poem by Rabindranath Tagore that became Bangladesh's national anthem, was a significant departure from the pre-modern agrarian imagination of well-being. Medieval poet Bharat Chandra wished for his progeny to live in 'dudh and bhat' (milk and rice), now such imagery is replaced by that of 'sona' (gold).

If the invocation of a golden agrarian past was required to awaken a 'dying nation' to the importance of the countryside and agriculture, it was also placed within a complex reinterpretation of race, religion and caste. In an attempt to recover the agrarian self of the Hindus, Pitambar Sarkar wrote that all major saints since ancient times had considered the Brahmin, Khaitrya and Vaishya castes as dijati, that is, of the same social denomination, although they had designated the profession of agriculture specifically to the Vaishyas. The *Bhagwat Gita*, according to Pitambar, considered agriculture as a 'natural' profession for the Vaishya and something that the Brahmin themselves could take up at times of 'emergency' (apadkala). Parashar of *Kali Yuga*, however, believed that the Brahmins could also take up agriculture in non-emergency situations.[7] Those who were already in agriculture but not considered one of the high castes, such as the Haldhar, were also given due attention in the new agrarian schema.[8] Jadunath Majumdar wrote that the Aryans were meant for agricultural work and they developed caste only when they were settled. Noting that Hindu society had become joro (stale), he suggested that the Brahmins should embrace the Namashudras and other low castes in order to stop the process of decline – an act that appeared as defeat but in reality was victory in a broader political context.[9] Within the larger context of the nation, the Muslims were not entirely excluded. Pandit Rajanikanta wrote: 'Say brother "Allahu Akbar" [Allah is Great] we are thirty crore brothers all connected, no one is different from each other / plough on the shoulder, sickle in the hands come brothers / we Bangalis will sow gold in the gold-producing fields.'[10]

The prevailing argument was that all prejudices had to be discarded in favour of agriculture. Ramhari Bhatcharya, a contemporary author, rejected the idea that nothing productive came out of the bhadraloks' agricultural activities; on the contrary he felt that they had some specific advantages in this sector. He argued that while most peasants had no capital of their own and the crops they produced were all promised to the mahajan before the beginning of the sowing season, a bhadralok engaged in agricultural work would have capital and would be free of the mahajan. Ramhari went on to tell bhadralok youths how to cultivate different crops and plants and earn at least 15 to 20 rupees per month. In an attempt to illustrate his project, Ramhari narrated the methods and advantages of potato cultivation through the words of a potato itself which, in its 'autobiography', rebuked the young bhadralok for their excessive focus on (government) employment and questioned whether their obsession with servitude to others would help them get swaraj. Referring to the enormous possibilities that lay in the cultivation of potatoes, it went on to demand: 'Since I have started my story I will not stop until I finish and make you listen in full. I will get you out of the obsession [with jobs] by any means. If I cannot do it myself, I will do so by the help of my other brothers [plants] and get you out of the obsession.'[11]

The invocation of a prosperous agrarian past was integrally related to the question of a possible territorial domain of productivity. If the Swadeshi existed in the countryside, it had to exist in a countryside that was capable of catering to the material needs of the 'nation'. By the turn of the twentieth century, ecological conditions in Bengal were on the decline in general, but western Bengal had been hard hit from the early nineteenth century. Dying river systems and the ravages of malaria had rendered a vast portion of the region as unproductive and largely dependent on the produce of East Bengal.[12] By the 1920s the situation was such that the director of agriculture of Bengal thought that 'the monotonous paddy cultivation appears unnecessary, at least in the rolling highlands of Birbhum, Bankura and Midnapore'. Instead he suggested that following proper attention to terracing and drainage it could be feasible within this area to 'render possible the production of cotton, pulse, sugarcane or fodder crops'.[13] No wonder that in most cases, the specific examples of 'Sonar Bangla' within the nationalist imagination were drawn from East Bengal. Brindabon Putatunda told a story of a hundred years ago in a remote village in Barisal called Kajlakathi, where a marriage ceremony was about to be spoiled by the sticky mud of the deltaic landscape. To save the situation, the host opened a few gola (large rice stores) to spread a layer of rice over the mud for the convenience of the guests. Putatunda lamented that the successors of that host were now plunged in such poverty that one couldn't but shed tears seeing their efforts to procure only a maund of rice for bare survival.[14]

The imagination of a prosperous East Bengal was linked to the appreciation of the ecological potential of the region. Writing in 1927, one Rishikesh Sen observed with contempt that the urgent call for development in the country – be it financial, educational or religious – only related to the so-called bhadralok. Sen observed that the rate of bhadralok unemployment was increasing and that they were growing in numbers, unlike the peasants and craftsmen whose numbers were decreasing. Therefore, as if to fill the vacuum created by nature, the bhadra santaenra (the children of the bhadralok) were taking up agriculture and craftsmanship. But Sen, concerned about the unemployment of the 'surplus' bhadralok, attached importance to establishing colonies in those parts of the country where vast wastelands existed. He explained that the practice of providing accommodation and employment by the methods of colonization within the country – known as 'home colonization' – was present everywhere in the world and that the establishment of colonies on the newly discovered islands was in fact the extension of this home colonization. Sen argued that since the colonial government would take them to new external colonies as 'coolies rather than colonizers', they should, for the time being, leave the thought of establishing colonies abroad and start settlements in uncultivated wastelands inside the country. The middle-class educated bhadralok, according to Sen, had to be the pioneers in agricultural initiatives in these colonies. Sen wrote that because of unemployment many bhadralok had taken up agricultural work, which was a 'very good omen', but he argued that though some bhadraloks were being employed, existing peasants were becoming proportionately unemployed, their lands being alienated. In this context too, he felt the need to create new employment for the landless peasants by the cultivation of new lands.[15]

From an ecological point of view, a striking feature of the early twentieth-century history of agrarian Bengal was the unprecedented influx of upper caste Hindus, namely, Brahmin, Kayastha and Vasyas, who together formed the bhadralok class in the active deltaic districts of eastern Bengal.[16] As late as the 1870s, Henry Beveridge was surprised by the absence of upper-class Hindus and Muslims in the coastal district of Barisal, for instance. Until about the turn of the century it was the West Bengal district of Burdwan and Kolkata city and the suburbs that were the popular home of the bhadralok. On the other hand, they were present only in some urban pockets in eastern Bengal. By the start of the twentieth century however the situation had changed considerably. Between about 1890 and 1930, the bhadralok grew in number more quickly in eastern Bengal than in western Bengal. While in the moribund western deltaic divisions of Burdwan and Presidency, the bhadralok grew by 17 per cent and 42 per cent (Kolkata included) respectively, in the active deltaic divisions of eastern Bengal, that is, Rajshahi, Dhaka and Chittagong, their number expanded by 53 per cent, 34 per cent and 93 per cent respectively. It was reported in the Census of Bengal in 1931 that in Chittagong almost one-half and in Noakhali nearly one-fifth of the total Hindu population were Kayasthas. In every other district of eastern Bengal, except in the Chittagong Hill tracts, they constituted

Table 5.1   Bhadralok population in the Bengal Delta

Division	1901	1911	1921	1931
Burdwan	634,175	674,031	658,352	746,632
Presidency	565,066	614,038	696,886	807,410
Rajshahi	133,543	153,277	180,195	205,355
Dhaka	608,653	655,232	769,907	861,696
Chittagong	264,925	302,363	384,865	511,325

Note: Data is extrapolated from data on caste as stated in the census reports on Bengal. The bhadralok, according to the census reports, represented the three upper categories of the Hindu caste, that is, Brahmin, Kayasthas and Vaishyas, who possessed immense political and social authority over the lowest and largest category, the Sudras. In particular, in 1931, 75 per cent of total Kayasthas lived in eastern Bengal. See also S.P. Chatterjee, *Bengal in Maps* (Calcutta, 1949), p. 54.
Sources: Census of India (Bengal), 1901, 1911, 1921, 1931.

at least 11 per cent of the total Hindu population, 'a proportion not reached elsewhere except in Calcutta' (see Table 5.1).[17]

The remarkable mobilization of the bhadralok in agriculture in eastern Bengal was evidently informed by the ecological condition of the region, which was better than that of western Bengal and Bihar. It was made possible by a number of evolving strategies, one of them being the mounting political pressure on the colonial government of the Swadeshi Movement. The following section focuses on aspects of the bhadralok's political engagement with the colonial administration in their bid to access the agrarian domain.

The agrarian bhadralok and the British Raj

The partition of Bengal in 1905 and its annulment six years later placed the Raj in a political dilemma. While the Muslim nationalist politicians of East Bengal considered the administrative-territorial rearrangement through the partition as an opportunity for self-assertion in a wider social and economic realm,[18] the bhadralok viewed it as a conspiracy to politically weaken the Bangali nation. Writing from South Africa in 1905, Mahatma Gandhi, referring to the news that 20,000 people had gathered in Dhaka to offer thanksgiving prayers to God for saving them from 'Hindu oppression', noted that even the founders of the British empire in India would have been opposed to the idea of partitioning Bengal, if they were to rise from their graves.[19] When the partition was revoked six years later, East Bengal Muslim politicians used the same language to express their anger. An English ICS officer, who viewed the revocation of the partition as an 'abject surrender to terrorism' and in some ways 'the beginning of the end of the British empire in India', referred to a draft speech of Nawab Salimullah. The draft, written by A.K. Fazlul Huq in 1912, declared that if the revocation was the King-Emperor's command, they must obey, but it added that 'Never again can the Englishman in the East put his hand upon his heart and say "This is my word".'[20] Rather than pacifying the bhadralok, the revocation of the partition encouraged the anti-partitionists to further strengthen their opposition to an apparently frustrated Raj. The process of partition and revocation thus resulted in Hindu and Muslim protagonists pulling the British Raj from two opposing sides.

In response, the government implemented a policy of engagement that was tailored according to differential communal moods. The Muslims, who thought they were not well represented in the government services and different representative bodies, were given particular attention. They were promised a university in Dhaka and other concessions and quotas. While the Muslim elite and middle-class protagonists measured their backwardness in terms of inadequate representation in government jobs and education as well as in electoral politics and wanted to catch up with the bhadralok, a large body of the bhadralok sought repatriation to the agrarian world of Bengal. These are particularly revealing developments in which an influential section of Muslims sought to take the positions which the general rank of the bhadralok considered materially inadequate. The government promoted these developments in the hope of taking the sting out of anti-colonial politics.

As mentioned earlier, throughout the nineteenth century, khas mahals or government lands were allocated to actual cultivators at low revenue rates and with special facilities in the deltaic regions of eastern Bengal. By the beginning of the twentieth century, this policy was compounded by a new policy of allocating these lands preferentially to the bhadralok. For instance, in the district of Barisal, of 36,004 acres of land settled between 1920 and 1930, the majority was settled by different categories of bhadralok.[21] In 1920, 330 acres in the Char Fasson estate in the Bhola subdivision was settled by members of the bhadralok community.[22] A medical practitioner was promised a huge amount of land in one of the blocks of char in Kukri Mukri on the condition that he would keep a stock of medicine as soon as the 'colonists' settled there after reclamation.[23] A witness informed the Provincial Banking Enquiry Committee that in every village of the Khulna district, the best agricultural lands were possessed by the bhadralok class.[24] In Noakhali, about 1360 acres of char lands were settled with three bhadralok youth societies.[25] A large tract of the Sundarban forest was leased to a young men's zamindary society which attempted to 'introduce cooperative methods in the cultivation of large areas and in the acquisition of zamindaries'.[26]

A scheme dealing with the question of unemployment among the middle-class bhadralok was initiated in the Faridpur district in 1928. The magistrate-cum-collector observed that the number of educated job applicants was far in excess of the posts available while 'the inability of Indian gentlemen of good position to find any sort of employment for their sons' was 'positively pathetic'. 'Back to the land', noted the magistrate, was one solution to the problem, but it was a solution to which both parents and young men were averse, and which would not be adopted voluntarily without 'substantial inducement'. The magistrate suggested that parts of the extensive chars or alluvial formations in the khas mahals of Faridpur should be made available for settlement in 'small, but remunerative quantities' to selected boys from the bhadralok class.[27] Accordingly, five young bhadraloks were trained annually at the Faridpur Agricultural Farm. After completion of their one-year training, each of them was given 15 bighas of khas mahal land rent-free for the first three years with no salami for zamindars being charged. They were also given an advance of 200 rupees in cash to buy a pair of bulls and the necessary tools. By 1933, the first two groups of youths had been provided with 150 bighas of land in one block in the Madaripur subdivision of Barisal, and the other two batches were given 200 bighas in one block in the Goalundo subdivision in Faridpur. It was noted in relation to the scheme that:

> The boys work in the fields just as ordinary labourers do, bare-footed and bare-bodied; and they prepare their own meals which consist of rice, *dal* (pulse) and fish which they get in plenty from the river flowing by their land. The local cultivators call them 'Volunteers', and some say that they are 'Swadeshi' people interned for committing offences against the State.[28]

In the 1916 edition of the *Bengal Waste Lands Manual*, it was stated that in cases where a bhadralok had undergone or was willing to undergo a course of agricultural training, or in cases where a bhadralok with a good working knowledge of agriculture was willing and able to undertake manual labour like an 'ordinary substantial cultivator', the collector was authorized to make a settlement of wastelands with him. It was proposed that the settlement would be made on the usual cultivating terms, but with a loan so that the bhadralok could reclaim and develop his holding.[29] Sixteen years later, the 1932 edition of the manual emphasized that in the khas mahals the primary object of the administration was to settle and maintain cultivating tenants on self-supporting holdings. Although the manual suggested that the suitable size of a holding might be five acres, it remarked that it was 'better to settle fewer people and keep them contented for a greater length of time than to settle more people and have them speedily discontented'. The manual suggested that settlements should ordinarily be made with the cultivating raiyats but when a bhadralok was willing to undergo agricultural training and to farm the land himself 'he might be admitted to settlement as an ordinary cultivating raiyat on a non-occupancy raiyati kabuliyat for not more than 10 years'. However, it was made clear that the area to be given to a bhadralok settler should be 'larger than that of an ordinary cultivator'.[30] In the 1936 edition of the manual, it was stipulated that the bhadraloks who decided to stay on the land and cultivate by themselves might be given settlement of holdings 'sufficient to enable them to live in a better style than ordinary cultivators'.[31]

Such preferential treatment was not entirely surprising because the avowed policy of the government was that of 'enabling persons who have abandoned terrorism to settle down to a life of productive citizenship'.[32] Even the apparently objective and institutional process of survey and settlement operations relating to land and revenue was considered useful in terms of the benefit of the bhadralok professionals. Regarding the necessity of an up-to-date record of rights, a government report noted that from the perspective of bhadralok employment, such proceedings were important since the greater part of the cost of revision operations, amounting to about 17 lakhs a year, went on employees' salaries, the majority of whom were bhadralok. It was thus considered 'an important asset to Government in easing the economic situation'.[33]

The open policy of preferential treatment of the bhadralok, however, had its problems. Khas mahals under government control were not unlimited. The government also had to be cautious in extending concessions to the non-cultivating bhadralok at a time when plenty of landless peasants were aspiring for quality lands and peasant unrest was growing apace. Therefore, the appeasement policy of the government towards the bhadralok may not be taken as a complete index of the complex agrarian relations in the first half of the twentieth century. We need also to look at the deep-rooted process of change in agrarian relations. This requires us to examine the hotly debated issues of land alienation and the shift of occupancy rights from actual cultivators to non-agricultural people. We will turn to these issues in the following two sections.

The bhadralok as raiyat: land transfer and the rise of neo-occupancy peasants

One strand of argument about the practice of land transfer in India is that, before colonial rule, landed property was never sold 'by order of a judge for payment of the debts of the owner ... voluntary alienation was allowed gradually, but the execution of decrees against a debtor's immovable' property was a thing not dreamed of'.[34] The East India Company introduced the system of land sale in case of default, resulting in the collapse of many zamindaris. Since a somewhat different pattern of land settlement developed in eastern Bengal, where the actual culti-vators were encouraged to occupy land rather than it being transferred, the legal question of transferability of land or occupancy rights did not become a major issue during most of the nineteenth century.[35] Referring to the Tenancy Act X of 1859, through which occupancy rights were legally secured to the cultivators who lived on a plot of land for 12 years, a zamindar commented that the right of occupancy was not transfer-able and that the raiyat could not transfer it by mortgage, sale, gift or exchange, nor was the right saleable in the execution of a decree for money against him.[36] The Tenancy (amendment) Bill of 1883 proposed to confer the free right of transfer, but the proposal was dropped in the face of opposition from the zamindars who were apprehensive of the growth of non-resident and non-cultivating raiyat.[37]

Yet the Bengal Tenancy Act of 1885 contained the seeds of land transfer as it left the question of transferability open to local customs, rather than to a prohibitive regulation. After the introduction of the act, the zamind-ars initially maintained an ambiguous position in relation to free rights of land transfer. The East Bengal Landholders' Association pointed out that the act of 1885 'encourages transfer, therefore they are bad. They hinder transfer, therefore they are bad'.[38] The zamindars took this ambivalent position because if lands were transferred from smallholders to undesir-able and financially competitive newcomers, their own position might have been compromised. The Bengal Landholders' Association noted that the land-dealers, mahajans and planters were the parties most interested in making occupancy rights transferable and it cautioned that if one wanted to ruin occupancy raiyats, one only had to give them the power of transferring their lands. It argued that it was possible to add value to a piece of land by attaching the right of transfer to it, but its value was never so great as when the cultivator retained the land and obtained a good crop. The association further noted that, 'by making it more valu-able in the market you tempt men to turn it into money, and transfer its value to other people. By preventing its transfer, you oblige him to keep its value for himself and his family'.[39]

While the zamindars were cautious in their response to attempts to legalize the right of transfer, they were equally aware that if land transfer was not legally sanctioned, they could not obtain a fee based on the transfer, thereby losing an opportunity for extra income. It is in this context that some leading zamindars pointed out that as far as the registration fees for the transfer were concerned, a fee of 25 per cent 'may not be deemed altogether inequitable'.[40] The 1928 Tenancy (Amendment) Bill confirmed that all occupancy holdings were legally transferable, subject to the payment of a landlord's fee which amounted to 20 per cent of the sale money, or up to five times the annual rent of the holding, whichever was greater.[41] Not surprisingly, it did not take long for the Bengal zamindars to make up their mind on this. They were also securing their place in the mahajani-credit market at this time, which fitted well with the growing trend of land transfer from occupancy raiyats to non-cultivating 'agriculturists'.[42] On the same ground, the zamindars 'did not exercise their veto on transfers to prevent genuine cultivators selling their lands to people who are not cultivators'.[43] Between 1885 and 1928, land legislation gradually tended to crystallize around the question of transfer and the process culminated with the introduction of the act of 1939.[44] The process of legalization of land transfer simultaneously led to the loss of land and occupancy rights to the actual cultivators and the impact of these developments was mostly felt in the comparatively fertile region of eastern Bengal, as Table 5.2 shows.[45]

Table 5.2, drawn from Partha Chatterjee's data, is especially important in terms of the propensity of land transfer in the active deltaic regions of East Bengal. Based on this data, Chatterjee, in keeping with many contemporary historians of Bengal, argues that the transferees of these lands were 'rich peasants'. This is a correct but incomplete diagnosis. In determining the real transferees of these lands, one has to be cautious about the term 'peasant' or 'raiyat'. In 1923, a cultivators' association of Noakhali complained that the 'greatest mischief' which was ever done to the cultivating classes of eastern Bengal was the enactment of certain sections of the Bengal Tenancy Act of 1885. The association contended that after the passing of the act, moneylenders and landlords, who were mostly non-agriculturist bhadralok, were allowed to take possession of most of the old raiyati holdings. The bona fide raiyats were compelled to take under-raiyati leases under those neo-raiyats who had now obtained the occupancy rights of the original raiyats.[46] It was noted that landlords generally would not allow anyone under them to have the status

Table 5.2 Percentage of occupancy holdings in each district transferred by registered sales, 1930–38

Western and Northern Bengal		Eastern Bengal	
District	Percentage	District	Percentage
Bogra	20.92 (partially active deltaic)	Noakhali	36.33
Midnapur	16.86	Tippera	30.34
Burdwan	14.99	Dhaka	20.92
Hooghly	12.72	Mymensingh	20.47
Howrah	12.25	Pabna	19.22
Birbhum	12.19	Faridpur	11.87
Nadia	10.04 (partially active deltaic)	Barisal	9.97
Malda	8.60	Khulna	5.03
Murshidabad	8.56	Chittagong	4.69
24-Parganas	8.25	Rajshahi	0.37
Rangpur	7.33		
Bankura	7.16		
Dinajpur	4.96		
Jalpaiguri	4.09		
Jessore	2.42		

Source: Partha Chatterjee, *The Present History of West Bengal* (Delhi, 1997), p. 62.

of more than a raiyat and that these bhadralok were 'only too willing to assume the lower status for that is very useful for them in ousting the under-tenants'.[47] One of the reasons for such a development, according to the association, was that under the Bengal Tenancy Act the under-raiyat was not given a chance to bid for the decretal amount in order to purchase the holding for himself. He was neither given the opportunity of paying the arrears in full nor of providing reasonable compensation to the auction-purchaser or decree-holder in order to regain control of his own land.

Section 104 of the Bengal Tenancy Act led to 'certain difficulties' involving the status of the bhadralok. The collector of Dhaka felt that by this provision the non-cultivators who sublet their land on a produce rent (barga) would succeed in being recognized as raiyats.[48] In 1914, the collector of Barisal reported that the general desire of all transferees in Barisal was to be classed as 'raiyats'. Indeed, even if they sublet all their land to another on the barga system, they would 'still be entitled to be recognized as raiyats'. The collector therefore proposed that the provision in the draft bill (amending the 1885 act) should be so changed as to render a transferee who sublet his land on the barga system liable to be classed not as a raiyat but as a tenure holder. For this purpose, the collector proposed that either 'subletting' be properly defined or the definition of the 'raiyat' in the Bengal Tenancy Act be amended so as not to admit a non-cultivator being classed as 'raiyat'.[49] A settlement report pointed out that though the Bengal Tenancy Act intended to protect the actual cultivator, it failed to do so primarily because both landlords and the courts treated the middlemen as raiyats and their tenants as under-raiyats.[50] The remarkable trend in the seizure of actual occupancy raiyat's long-accumulated rights and entitlement was illustrated by a settlement officer from Chittagong who remarked during the Depression era:

> Formerly the term *chasa* (cultivator) used to stink in the nose of the cultivators themselves as being a term of opprobrium. But at the advent of the present century the gradually increasing unemployment among the middle class youths and the prevailing high prices of agricultural produce brought about a change in the out-look and while a member of middle-class took to agriculture as the only available profession others were tempted to lay out money on lands as a form of safe and lucrative investment ... In the world of trade, atmosphere of doubt and uncertainty ... those who have money to invest are acquiring landed property as being the safest mode of investment. This coupled with the comparative opulence of the non-cultivating classes has certainly accelerated the process of transfer of lands from the cultivators to the non-cultivating classes ... the transfers count among themselves not only the money-lenders, but also landlords and pleaders, muktears and service holders ... A section of the transferees let out their lands on Bhag or Barga system and take half the produce as rent while they let out on a high rate of money rent.[51]

Although the process of land transfer and relocation of occupancy rights meant the collapse of the agrarian economy as sustained by the original occupancy raiyats who had emerged in the course of the nineteenth century, this should not have in itself caused remarkable displacement in the existing agrarian production process, since the bhadralok as a neo-occupancy raiyat was under the impulse of a new nationalist discourse of rural development, patriotism and self-help. But the bhadralok's assumption of occupancy rights did not generate much of the prophesized regeneration of rural Bengal. On the contrary, the bhadralok's involvement with the countryside coincided with a period of social and economic decline that culminated in the great Bengal famine of 1943. In order to examine the failure to translate the nationalist imagination of collective well-being into a reality, we need first to study how the bhadralok negotiated their new-found rights and access to land and to the social production process in the context of a corresponding displacement of the actual cultivators.

The nation and its bargadars

The imperative for the bhadralok to return to the agrarian milieu was associated with a variety of opinions on how the renewed engagement in agrarian production processes could be fashioned.[52] Difference of opinion largely emanated from the aversion of the bhadralok to manual work.[53] Some bhadraloks, while fully endorsing the idea of a retreat to the agricultural arena, argued about the question of physical involvement in cultivation. Addressing both 'less' and 'highly' educated youth, Biseshwar Bhatcharya noted that 'mother Bengal' expected a lot from them and that they should start working for themselves and the country by toiling in the countryside. He advised them not to be afraid of malaria in the villages and not to think that life without electric lights and bioscopes was worthless. Biseshwar, however, emphasized that no one was advocating the 'wild idea' that educated young bhadraloks from towns would immediately take up the plough. Biseshwar asserted that even without the bhadralok undertaking their own cultivation, something could still be done for the economic well-being of the country. He felt that there were 'enough people out there to handle ploughs', such as bargadars. What was important for a bhadralok, according to Biseshwar, was to employ his knowledge, education and goodwill in order to work for the betterment of himself and his fellow countrymen.[54]

Another writer complained that those who advised the bhadralok to engage in cultivation and other such physical labour were unaware of the fact that if bhadra (gentle folk) did such work, no work would be left for the itor (non-bhadralok), and that the itors were better off only because the bhadras did not do all such works, not because abhadra (non-bhadralok) were skilled workers. Second, he argued that most village-dwelling bhadralok had land, but that agriculture was decreasing due to want of cultivators. A remunerative agricultural holding could not be the result of one single person's physical labour. Third, the author argued that the bhadras had not done agricultural work for generations and that if they were to do so now their social respectability would be lost. He added that the desire to protect and promote one's honour was moving people not only in Bengal or in India, but all over the world and that attempts to become saints by suppressing these desire would require the whole universe to be reordered. He asked empathetically whether there was no elite in *Bilat* (England) or whether there was no difference between gentlemen and shopkeepers there? He added that the son of an American president could carry things on his head because there were no Brahmans or shudros, no lords or ladies, hence there was no point in raising the issue of America when talking about people carrying things on their heads in India. He went on, 'we kept the school boys in the beautiful *bilati* gardens, now telling them to come out; we taught them *bilati motigoti* [English attitudes and approaches], now telling them not to cut *teri* (stylish haircuts), or wear socks. He asked how he who had not done any physical work for sustenance for many years will adapt now?'[55]

The dilemma faced by the bhadralok in making agriculture their prime source of livelihood while avoiding direct physical involvement in the agrarian production process was largely resolved by the bargadars and landless cultivators who emerged in the wake of the land transfer spree from the actual occupancy cultivators to non-agricultural landowners. By obtaining the land and occupancy rights of the raiyat through both legal and coercive means, the bhadralok acquired, ready-made, the facilities and privileges that had accrued to the peasant over the course of the nineteenth century. The number of original cultivators possessing occupancy rights had been undergoing 'steady reduction' since the beginning of the twentieth century and between 1921 and 1931 there was a 49 per cent increase in the number of landless labourers. By 1940, landless labourers constituted 29 per cent of the total population in Bengal.[56] Shantipriya Basu observed that zamindars and majahans were buying the cultivators' lands and returning them in barga to those cultivators, since it was more profitable if land was first made 'khas' and then let out in barga. Following the data on a sample survey reported in the Floud Commission Report, Basu noted that about two-thirds of the land sold was cultivated by people other than the buyers.[57]

As the bhadralok became de jure raiyat, they opposed any proposal that entitlement might be offered (or returned) to the under-raiyat or bargadars. One argument from the bhadralok about retaining the share-cropping arrangement was humanitarian in nature. The joint secretary to Bengal Mahajan Sabha noted that a large number of poor and middle-class families who could not cultivate their land for themselves had it worked by sharecroppers. He argued that these families often consisted of widows, minors and the disabled, who might be deprived of their means of subsistence if the system of barga was discontinued.[58] A pleader from Comilla advised that the occupancy right should not be granted to under-raiyats and bargadars, and that 'raiyats' should not be allowed to grant leases for indefinite periods as it would result in lands going out of their control.[59] A pleader from Rajshahi argued that a landless class of labourers would always exist in the country and that he found 'nothing inequitable' in a barga arrangement between members of this class and the 'middle-class raiyats'.[60] A zamindar observed, on behalf of the Landholders' Association of Barisal, that the bhadraloks who were not accustomed to handling ploughs themselves were in the habit of extensively settling their lands in barga for 'fear of loss of honour and dignity or for a variety of other reasons'. He argued that the share that the bhadralok received from the bargadar as the landlord's portion of the produce constituted the main source of subsistence of himself and his family. He felt that if the bargadar was given occupancy rights he would have recourse to 'convenient expedient of commutation, depriving thereby the landlord of his portion of the produce on which a money value will be put falling far short of its market value'. The consequence, he continued, would mean 'complete effacement of the mediocre gentry from the social economy of the country at no distant date'.[61]

The argument for retaining the status quo on humanitarian grounds was often supported by the plea that the presence of the bhadralok was necessary in the countryside on account of their intellectual superiority. It was suggested that if the bhadraloks were deprived of the facilities of barga system, they would lose interest in the villages and migrate to towns. Through this process, the villagers would be 'deprived of their intelligentsia' and at the same time the moneylenders would take advantage of the situation.[62] Such arguments were bolstered by the delicately phrased threat of political instability. In the Dhaka district, for instance, it was argued that if the bargadar acquired the status of a 'tenant', the owners of these lands would 'almost certainly cease to let them out in this way' and that the land would be left fallow benefiting no one and leading to rising unemployment and unrest.[63] A munsif to the district judge of Rajshahi wrote that a large number of clerks had been working on very low salaries only because most of them had their small stock of grains at home to depend on. The munsif warned that if they were deprived of this supplement to their wages the establishment would have to 'face a community of educated and discontented people and lawlessness started by them would be more difficult to deal with than the disorganized out-burst of ignorant criminals'.[64]

The staunch resistance by the bhadralok to the restoration of any entitlement to the bargadar was consistent with the process of the transfer of land and occupancy rights from the cultivators. The process would have lost all meaning for the bhadralok if the sharecroppers and landless were legally empowered to regain their rights and entitlements to land. This necessitated strong measures to protect the agrarian space of the new generation of non-agricultural occupancy raiyats against any bid to restore the rights of the bargadars. Quoting Manu Samhita, Radharomon Mukherjee, a contemporary expert on land law, believed that 'as the wild deer of the forests became the property of the hunter, so did the arable land become the property of the man who first reclaimed the forests for cultivation'.[65] Radharaman believed that holding even under bhagdari (bargadari) tenure would establish a right of occupancy. But he suggested that this Puranic legal code did not apply to East Bengal where 'the local custom recognizes in such cases no right to the land in the cultivator, but merely to a share of the produce raised by him'. The bargadar in the district of Pabna, for instance, was ordinarily a cultivator who, under the terms of the contract, was a 'servant or labourer under the holder of the land'. Mookerjee also believed that the accrual of occupancy rights did not depend on the raiyat's cultivation of land by himself and that the law did not require that a person in order to acquire the right (of occupancy) should be a bona-fide cultivator.[66]

Observing the policy shift towards conferring rights of land transfer on the raiyats, in 1905 an inspector general of police for the Lower Province suggested that it would be impossible to keep out the middleman and the mahajan if the right of transfer was granted to the raiyat.[67] By 1915 legal amendments began to validate the transfer of an occupancy-holding to 'a person who is not a cultivating raiyat or has not acquired such holding ... for the purpose of cultivating it by himself or by members its family or by hired servants, or with the aid of partners'.[68] The process of land transfer and sharecropping that simultaneously developed after the introduction of the BTA culminated in its 1928 amendment, which accommodated the key demands of the bhadralok. The act consolidated the existing rights of the 'raiyat' and denied any entitlement to land on the part of the sharecropper. A decade later, in 1938, the act was modified again and the rights of the 'raiyat' were further fortified; at the same time the last constraints on land transfer were lifted. The new act also provided that no increases in rent would be allowed in the following ten years. Ironically, this reform meant hardly anything to those actual cultivators who had become bargadar, under-raiyats or had joined the ranks of the landless and thus had found themselves ejected from the legal domain of raiyathood. The Bengal famine of 1943 took its toll in particular from these luckless victims of a changing agrarian society. Although historians have been correct in recognizing the 'rich peasant' as a key beneficiary of agrarian resources, they have remained reluctant to identify the shifting identity and role of a new agency that occupied the semantic, legal and social space of the category of the 'peasant'.

Agrarian relations between 'world capitalism' and micro-credit

The initiation and sustaining of the process of land transfer and share-cropping was not only the outcome of a series of political and legal measures aimed to secure landed property and associated occupancy rights for the bhadraloks. While government policies and tenancy acts favoured land transfer and sharecropping, the trend increasingly appeared to be related to and accentuated by an epidemic of indebtedness.[69] According to the 1933 report of the Bengal Board of Economic Enquiry, the average debt per family in Bengal increased at least threefold between 1906 and 1933 and, in the same period, the proportion of debtless families decreased from 55 to 17 per cent (Table 5.3).[70]

It is difficult to gain a clear insight into the particular agency of any creditor in late colonial Bengal since a remarkable convergence among creditors of different backgrounds was taking place.[71] The involvement of traditional mahajans and post-depression era 'rich peasants' is narrated in detail in the works of Sugata Bose. By the beginning of the twentieth century a section of the bhadralok had joined the ranks of traditional mahajans and the landlord-cum-moneylenders.[72] In Mymensingh, for instance, professional moneylenders and merchants-turned-moneylenders provided around 50 per cent of loans made to agriculturists. Some 30 per cent of loans were believed to have been obtained from the 'richer agriculturists or land-holders or men of

Table 5.3   Average debt per family in Bengal, 1906–33 (rupees)

Estimate	Board of Economic Enquiry 1933	Mr Barruge's report 1929	Mr Jack's report 1906–10
Average debt per family	217	128	55
Average debt per indebted family	262	203	121
Percentage of debtless family	17.0	37.0	55

Source: Shantipriya Basu, *Banglar Chashi* (Calcutta, 1944), pp. 26–7.

other professions who conducted money lending as a supplementary profession' who were taking the place of professional money lenders – the shahas, banikyas and telees.[73] Government clerks and even widows took up money-lending as a subsidiary profession.[74]

Agrarian credit relations in Bengal have been the subject of a number of important works. To avoid repetition, this section will examine cooperative credit, a particular type of rural credit mechanism that was considered an ideal institutional alternative to the coercive practices of traditional mahajans.[75] First conceived in France and then extended by Robert Owen in early nineteenth-century England, the idea of coopera-tion between the working classes and peasants in Europe and in the USA remained very popular and travelled to many parts of the colonial world. In the wake of an emergent global cooperative movement, the first Co-operative Societies Act (Act X) was passed in Bengal in 1904. One of the aims of the movement was to encourage investment in those parts of the country where there was possibility of activating 'idle capital' and eastern Bengal appeared to be an ideal region for achieving this objective. It was reported that within five years of the start of the move-ment there was evidence of ample collection of loans and deposit money along with the 'successful tapping of hoarded gold out of its conceal-ment' of which 'some particularly striking instances' were reported from eastern Bengal.[76] In 1935, among the 26 districts of greater Bengal, ten eastern districts hosted 10,749 credit societies out of the total of 20,054, and possessed Rs. 302.3 lakh (30.2 million) working capital out of the total Rs. 519.1 lakh (51.9 million).[77] Between the 1910s and the 1930s, the number of individual financiers rose from 230 to 5658, including influential public figures like Sir Ashutosh Mukherjee.

In the management of cooperative credit networks, the government of Bengal took the role of a supervisory authority while a three-tiered non-governmental institutional structure was established to oversee its overall performance. At the lowest level were the 'primary' or village societies, in the middle the 'central banks' and on the top the Bengal Provincial Co-operative Bank. The Provincial Bank worked as the apex body of all central banks and as a coordinator of capital flow between different central banks. The net cash profit of the Provincial Bank dur-ing 1938–39 amounted to Rs. 81 lakhs.[78] Central banks acted as inter-mediaries between the primary societies and the Provincial Bank and other private banking institutions in Calcutta. The main function of a central bank was to serve as a link between the societies and the money market, and 'a bridge spanning the gulf of mutual strangeness over which business can be carried backward and forwards'.[79]

A planned process of overhauling the 'idle capital' of East Bengal and a significant amount of working capital in the hands of the cooperative programme ushered in the possibility of self-help and an alternative way of utilizing rural capital in productive ways. This did not happen. On the contrary, during the operation of the cooperative movement the problems of indebtedness, poverty, lack of capital formation and low productivity increased. This failure demands an explanation. The real motivation for investment in these credit societies was related to the fact that these were of Raiffeisen type. The unlimited liability on which they operated implied 'a careful selection of members and a restricted area of operations'. Shares and dividends in the form of previous accumulation of capital were not considered 'suitable to the economic conditions of villages in Bengal' and profits therefore went to the reserve fund, which was indivisible.[80] In the wake of world economic depression it was further brought to the attention of potential investors that the Raiffeisen type was suitable for rural people who formed a compact and homogeneous group, socially and economically, and who therefore were 'in a position to accept individual and collective responsibility for one another's debt'.[81] It was expected that the cooperative societies would attract considerable amounts of capital within this safer framework of capital flow. It proved successful in this regard. Perhaps it is unsurprising that most of the capital of the cooperative credit societies was used as investment for profiteering. For instance, the Barisal Co-Operative Central Bank Limited opened hundi trading and advertised that anybody who deposited money in the bank could profit by sitting in Calcutta only at a minimal commission and that mahajans of Calcutta could obtain money without any hazards by accepting hundi against the Provincial Banks. This way, this particular bank did hundi karbar of Rs. 87,167 in 1928–29.[82] While the borrowing rate of the Bengal Provincial Co-Operative Bank varied from 4 to 5 per cent per annum, the final rate at which villagers got their loans from societies varied from 12 to 15 per cent.[83]

Given the stake of the influential bhadraloks in the cooperative regime, it was expected that 'the loan companies' would be excluded from the operation of the Bengal Agricultural Debtors Act on the bases of 'inconvenience', and that their inclusion would prove both 'troublesome' and 'expensive' as they were in remote rural areas.[84] The cooperative societies were also exempted from the operation of the Debt Settlement Boards in the late 1930s and early 1940s. A case dealing with cooperative debt could be kept pending for an indefinite period as no decision could be reached without the approval of the Co-operative Inspector, a process that could itself take an indefinite time.[85] There were official warnings against rushing cooperative debts through the mechanism of Debt Settlement Boards, and it was believed that where cooperative credits were included within the scope of debt-adjustment schemes, such debts were 'safeguarded' and that the cooperators had 'got about all they wanted by the back door of the Rules'.[86]

Rather than encouraging cooperative sale or production societies, the cooperative movement promoted the credit business and therefore became part of the problem which it aimed to eradicate. For instance, in 1923 in the United States the farmers' organizations did business worth two billion dollars. Of the 8913 organizations which submitted their reports, 90 per cent were engaged in selling farm products.[87] In Bengal neither the cooperative stakeholders nor the government made any special efforts in this direction. In 1925 there was one cooperative jute sale society, by 1926 there were ten, by 1929 the number had decreased again to 5, and such societies ceased to exist at all after that date.[88] These societies were financed for the purchase of jute by the Bengal Provincial Co-operative Bank Ltd of Calcutta on the recommendation of the Bengal Wholesale Society Ltd. They sold jute to the mills as well as to the European brokers and they got better prices than the local merchants. What is remarkable is that very few of the members of the societies were actual cultivators.[89] Where actual cultivators were able to form societies, they were not linked to any central organization, nor were they assisted by outside financiers: 'the nearest central bank of credit type is unwilling to finance such societies as their liability is limited by shares, which are hardly sufficient to justify for their requisite borrowings. They have, therefore, to depend on their share capital and local deposits, if any, to carry on their business and as such are generally handicapped by their finance.'[90]

A crucial factor for the rise and failure of the cooperative movement in colonial Bengal was the way it which it was linked to a project for the rehabilitation of the bhadralok. With the introduction of the Cooperative Societies Act the government of India requested the provincial governments and swadeshi leaders to establish cooperative societies.[91] Lord Curzon paid 'unusual attention' to the scheme of cooperative societies.[92] It was suggested that 'they [swadeshi leaders] should select competent leaders to promote agricultural and commercial undertakings, found agricultural associations, Co-operative societies [for the youths]'.[93] Considering the cooperative credit movement a 'revolutionary' idea, P.J. Hartog, the first vice-chancellor of the University of Dhaka, noted that there was a tendency in Bengal to make what he called a 'salary caste', and he hoped that the cooperative effort might effect a revolution in agriculture in Bengal as it had done in Denmark and Ireland. By uniting the peasantry of Bengal, the movement, Hartog thought, would boost both large and small-scale cultivation projects and pave the way for the employment of a large number of bhadralok.[94]

At both central bank and primary society levels, the bhadralok were given a free hand in the management of the cooperative societies. For example, the Barisal Central Bank gave additional loans to more than one person from the same family, and the bank chairman and the secretaries laundered (attoshat) the major share of the assets.[95] On the Sandwip and Hatia islands in around 1920, Rajendrakumar Naga established several central cooperative banks. In 1924, the total capital of these banks was Rs. 100,000. In these coastal areas 'many foreigners' had come and settled. In Sandwip these came 'with very little capital in hand and now have become very rich'. Meanwhile, three-quarters of the peasantry of Sandwip could not afford full meals twice a day regardless of the amount that they produced, and they could not die free of debt.[96] In the Barisal Sundarbans, the cooperative movement could not stop rural indebtedness and declining trends in the agrarian economy even though earlier arrangements for leasing lands in large blocks to capitalists were discontinued in order to enable direct dealing with the raiyats.[97] In the Khepupara colonization project in the Barisal Sundarbans, where all forest lands were reclaimed, the average raiyat's debt – comprising the rent, cess and government loans – stood at Rs. 100 whereas each member of the cooperative movement had loans of more than Rs. 200. This sharply contrasted with the bulk of the working capital of the central banks of the area, which provided loans to its 109 member primary societies of Rs. 6 lakh.[98]

Corruption and mismanagement caused the cooperative movement to fail from its inception as far as rural uplift was concerned. Niyogi observed that it would be a misreading of the history of the cooperative movement to suggest that the economic depression of the 1930s was chiefly responsible for its decline, for the seeds of decay were 'soon broadcast very early in its career by a disregard of those rules of prudent finance and efficient administration'.[99] The Depression years, however, introduced a new language and rhetoric to the movement. From a political point of view, the initial years of the movement saw government trying to use it to contain swadeshi unrest; later it was the terrorist movement and communism which informed the cooperative organizations. Alexander Hamilton, an enthusiastic cooperator, informed the Governor of Bengal that a cooperative utilization of merely 10,000 rupees could eventually bring a revolution in the economy of Bengal to such an extent that 6000 'White Shirts' (in the place of 'Red Shirts') of the great peace army will pave the way for 10,000 doctors, and 100,000 teachers, and 'unemployment of the bhadralokh will wither and die'. Hamilton recalled a conversation with a 'brilliant young product' of Calcutta University in which, while talking about nationalism, he said to him that the only 'ism' he bothered about was the 'belly'. Hamilton in this connection reminded the governor that if he could 'apply 50,000 bread poultices to the bellies of young Bengal, the fever of unrest will abate, and vanish eventually like an evil dream'.[100]

The government was apparently moved by Hamilton's plea and similar arguments by others. Governor Anderson, in a speech in 1935, reminded the bhadralok of the disadvantages of a 'perverted form of terrorism, and ... anarchy in the shape of non-cooperation and civil disobedience'. Among other pieces of advice, he asked them to return to their villages, to sit on the Union Boards and to plan constructively for the improvement of the rural areas and to take the lead in forming 'co-operative societies for a multitude of purposes which will band the people together in small units working for the mutual advantage of their members'.[101] Given the degree of politicization of the movement, it was no wonder that by 1940, far from wiping out rural debts the societies had become another 'incubus on the tenantry'.[102] Ian Catanach observed that in the Bombay Presidency there was little evidence of cooperative societies being used for political ends and that while the movement had by no means intended to extinguish the moneylender, it did aim 'to put some fairly definite limit on his activities'. Catanach noted that one of the aims of the cooperative movement in Bombay was to prevent land from being passed to moneylenders from agriculturists and that in this way it contributed to the general well-being of the peasants. In Bengal, the movement remained politically biased and less effective.[103]

To conclude this chapter, it may be argued that the return of the bhadralok to the agrarian domain of eastern Bengal was foreseeable in the context of a number of push factors such as urban unemployment and pull factors such as the allure of the last viable agro-ecological frontier of Bengal. But the return of the bhadralok came at a time when frontier fertile rice lands were already becoming subject of a scramble by a fast-growing peasant population. The Sundarbans forest system was cleared up to the sea in these fertile regions, especially in Noakhali and Barisal. In inland areas, such as Mymensingh, the alluvial char lands were also being contested as demographic pressure mounted. Extensive uses of land also caused fertility decline. At a time when – at the peak of ecological exhaustion through the nineteenth-century commercialization of agriculture – agrarian policy in eastern Bengal began to be characterized by intense contestations between demographic pressure and decreasing ecologically-endowed land resources, the return of the bhadralok into the agrarian production process was bound to complicate political, social and economic relations with far-reaching consequences – a subject that we will take up in the last chapter. The following two chapters will look in greater detail at the changing ecological conditions in which the agrarian and social changes, as described in this chapter, took place.

6
The Railways and the Water Regime

The railways in India drew considerable attention from two of the most influential thinkers of modern times, Karl Marx and Mahatma Gandhi. In the 1850s, Marx was a distant but passionate observer of the emergence of the railways in India and he was convinced that the new transport system would prepare the ground for a bourgeois civilization, precursor to socialist revolution. Apparently informed by the nineteenth-century spirit of 'improvement', Marx linked the railways to industrialization, communication and formative intercourse between the inhabitants of disparate villages, communities and castes across India. He also envisioned that the railways would lower the intensity of famine by mitigating the problem of means of exchange; and that the digging of tanks or borrow-pits for embanking the railways would lead to the creation of an extensive irrigation system which would contribute to agricultural development.[1] About half a century later, Mahatma Gandhi took a completely different view. He not only denounced the railway for its role in promoting British imperial penetration in India, but he also blamed it for the transmission of plague and the bringing of famine by draining lands that would otherwise have been cultivated in the hinterland of India. Gandhi also complained that the sanctity of holy places – whose inaccessibility had meant that they were visited only by real devotees prepared to endure difficult journeys –had been lost as a result of easy railway transportation and visits by 'rogues'.[2]

The apparently contradictory approaches of Marx and Gandhi towards the railways seem to have been informed by their different views of modernity. For Marx, modernization was imperative for India's dialectical advance towards materialistic progress; for Gandhi, modernity, or the 'disease' of civilization, destroyed the inherently rhythmic and sacred autonomy of traditions. Throughout the twentieth century the historiography of railways in India seems to have either endorsed or contested these early assumptions. For the followers and critics of Marx, the building of the railways has generally been interpreted as an imperial scheme designed to mobilize capital, raw materials and troops for better or for worse. Its impact has been measured by the macro-economic performance of India within the general pattern of industrialization, export and import.[3] On the other hand, a number of recent works have cast doubt on the modernization paradigm itself. A critique of the 'civilizing mission' suggested that railways were essentially imperial projects demonstrating the cultural and technological superiority of the West.[4] Other research in this arena relates to the ways in which railways have spread disease – and even communalism.

While these debates focus on political-economic or socio-cultural issues, there has been a remarkable lack of study of the railways in conjunction with the landscape and the environment through which they ran. This 'third line' of analysis began to emerge recently, in particular with the works of Ira Klein. Klein has convincingly revived the early twentieth-century arguments of C.A. Bentley, Director of Public Health in Bengal, that the railways caused ecological deterioration and consequently increased malaria and morbidity.[5] This chapter extends the debate on the links between the railway and disease to links between the railway and the larger water regime of the Bengal Delta. In doing so, I will examine two broadly related questions. First, in what ways did the concept of modernity as expressed in the development of the railways excluded environmental considerations? Second, how did the inevitable conflict between the demands of the railway and the water system of deltaic Bengal lead to considerable deformation of the landscape and the destruction of agrarian production processes? In engaging these two questions, I would like to map the discursive space within which 'modern' knowledge, colonial capital and Bengal's fluid landscape defined the railways.

The bottle and the funnel: the coming of the railways to the Bengal Delta

After conducting a year-long survey of landscape, possible routes and profitability, Macdonald Stephenson, a Scottish engineer, proposed the first Indian railway scheme in 1845. Stephenson's ambitious scheme of 'triangulating India with railway' – at an estimated cost of about fifty million pounds – envisaged the creation of a vast railway network that would connect Kolkata, Delhi, Mumbai and Madras and other major towns in between. Doubt was cast on the practicality of such a huge project, but the debates that followed endorsed the necessity of railways in India. As far as the territorial breadth of the railways was concerned, both Stephenson and his early critics envisioned Kolkata as the eastern-most terminal of the future railway network in India. Regions east of Kolkata or eastern Bengal, along with Assam, were excluded from the purview of the scheme.[6] (See Figure 6.1.)

The disadvantages of excluding eastern Bengal from the initial rail-way projects were soon identified. A fifteen-page monograph, published in 1848 by one 'Transit', pointed to the relative merits of extending the railways into the Ganga valley. He strongly criticized Stephenson's new East Indian Railway Company, which was to connect Kolkata to Bihar and then to run through the Doab region via the short but circuitous route of the Rajmahal Hills. Transit thought that the projected line ignored Bengal trade and would get 'out of Bengal as fast as it could into the hills'. In insisting that the trade of Bengal should be considered in

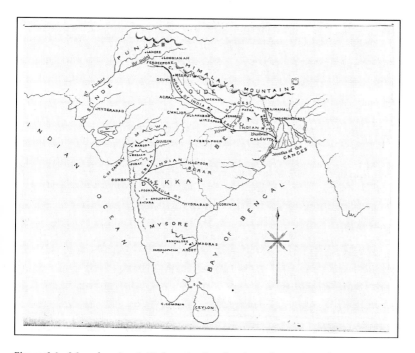

Figure 6.1 Map showing initial route plan for the railways in India
Source: An Old Indian Postmaster, *Indian Railways; as Connected with the Power, and Stability of the British Empire in the East* (London, 1846).

the future expansion of the railways in India, Transit appeared to have been informed by a wider vision of the water regime of the lower Ganga valley which provided the sole impetus for an extensive range of trade and commerce. Transit reminded the British and Indian capitalists:

> The Ganges Valley is your manufactory – your trading ground – your source of wealth. I look not to towns, to provinces, to districts, or to individuals; I look not to transporting sepoys, or cannon, or gunpowder, or arms ... not to Manchester twist, or Welsh iron, or Swansea copper, or French brandy, or Burton ale; I look not to Purneah indigo, Patna opium, Benares sugar, or Chuppar saltpeter, Mirzapore cotton, or the grain of the chete; but, on the broad principle of the greatest benefit of the greatest number, I say, that by the Ganges you catch the whole.[7]

Transit suggested that the resources of north-eastern India could be better served by connecting Kolkata by railway to the 'nearest permanent spot' on the bank of the lower Ganga Valley. He found the starting point of the Bengal Delta, where the Kosi river meets the Ganga near Malda, to be an ideal terminus for a railway from Kolkata. He argued that the whole accumulated trade of the Ganga valley, comprising an area of 150,000 square miles and containing a population of 40 million, was bound to pass through this 'narrow neck' of the country, not more than five miles in width. Transit compared the lower Ganga valley to a funnel whose apex was the starting point of the delta and he viewed Kolkata as a bottle which would draw the trade of the Ganga Valley through the funnel where the proposed railways would work like a pipe. Transit's scheme was significant in view of the proposed railway's contact with the highly fluvial landscape of the Bengal Delta. He did not consider railway routes further down the Ganga valley. Though the huge commercial prospect for the Bengal Delta was present in his mind, Transit did not even want the railways between Kolkata and the Kosi-Ganga bank to have branches: 'You should put a pipe to the apex of the funnel, and its lower end in the bottle, not to climb up the side and take a drink at the edge, or to make furtive hole in the side by which you will only drain it half-way.' In insisting on a single principal line, Transit seemed to have considered that the 'richest land would be found near the base of the drainage'. He considered the Ganga trade route of the pre-colonial period, but only to the point at which the trade flow could be diverted to Kolkata by train. Thus, according to Transit's scheme, though the resources and trade of all north-eastern India could be tapped, the physical presence of the railways into the 'base of drainage', or the Bengal Delta, was not proposed. Transit's idea of the railways as a means of tapping the resources of the delta was perhaps influenced by the strength of its water regime which ensured the transportation of enormous amounts of raw materials, trade and commerce. According to Transit's scheme the railways were supposed to have a complementary rather than a confrontational relationship with the deltaic water regime.

Other views, however, considered the Ganga as a competitor to the railway, and if the expansion of the railways in Bengal over the following decades is examined in relation to Transit's scheme, it would appear that while the commercial importance of the Bengal Delta as pointed out by Transit was fully taken into consideration, there was a fundamental difference between his scheme and the railway projects that were actually carried out. Instead of engaging with the Ganga Delta via the 'neck', the railways entered its fluvial heart, and in consequence they began to contest rather than complement the water regimes of the delta. With a view to eventually connecting Kolkata with Dhaka, the first railway line was opened from Kolkata to the lower Ganga bank in Kushtia in September 1862. In 1871, this line was extended southward to the Goalundo bank of the Ganga. With its many branches extending along both banks of the lower Ganga, it came to be known as the Eastern Bengal Railway (EBR). Between 1874 and 1879 the Northern Bengal State Railway, extending from Sara to Sirajganj, was constructed, with branches extending to Dinajpur in the west and Parbatipur in the east. In July 1884 the government acquired the EBR and in 1887 it was merged with the Northern Bengal State Railway. The entire Eastern Bengal Railway (the word 'State' was dropped in 1915) was situated on the west bank of the Brahmaputra river, with the single exception of the Bahadurabad-Dhaka-Narayanganj line. The first section of the Assam-Bengal Railway (ABR) was opened between Chittagong and Comilla in 1895. The line was constructed to meet the demands of the tea companies in Assam, which wanted railway facilities for the export of tea via the port of Chittagong. This line lay on the left bank of the Ganga and both banks of the Brahmaputra. It served the province of Assam, and the districts of Dhaka, Mymensingh, Chittagong, Noakhali and Comilla. In 1942, the ABR was taken over by the state and was merged with the EBR to form the Bengal and Assam Railway. The expansion of the railways was such that by 1933 Bengal had more railways by area than any province except the United Provinces.[8] (See Figures 6.2 and 6.3.)

The roots of the area's ecological problems appear to lie less in the construction of the railway itself than in the fact that the government

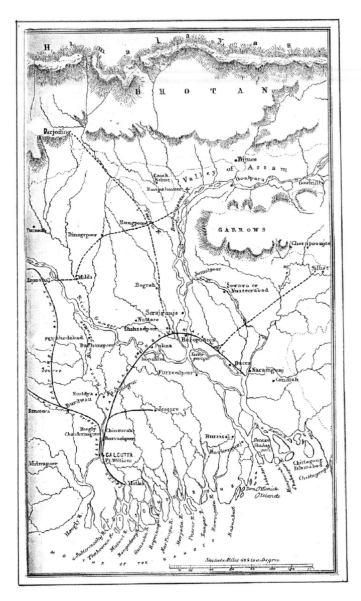

Figure 6.2 Map showing the emerging railway routes in eastern Bengal
Source: *A Sketch of Eastern Bengal with Reference to its Railways and Government Control* (Calcutta, 1861).

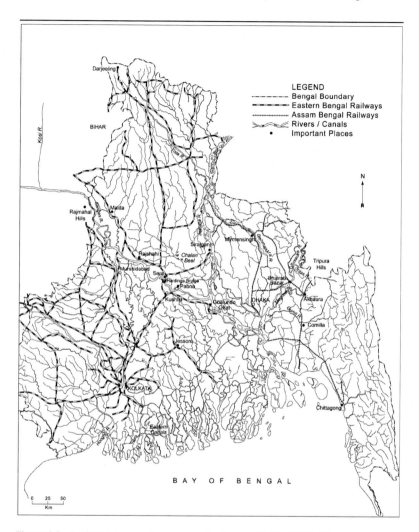

Figure 6.3 Railways and waterways in the Bengal Delta, 1938–39

and the different railway companies, while encouraging the construction of the railways, failed to appreciate the relative importance of inland waterways. In the Bengal Delta, waterways were often seen as rivals to the railways as means of transport and there was a feeling that with the completion of the railway networks, the 'slower' mode of transportation by water would be rendered obsolete by the new, faster and more reliable transport and communication system. The question of investment in the railways in eastern Bengal was brought to the attention of the government as early as 1852, well before the experimental line of the East Indian Railway Company from Kolkata to Raniganj coalfield was tried out. In 1856, the merit of the rail line project, between Kolkata and Dhaka via Jessore, was 'tested' on the London financial market and the enthusiasm with which the shares were bought was 'perfectly astonishing'. The capital for the first section of the line was estimated at £1 million Sterling, but the applications actually made were worth over £15 million. By the end of the 1850s the EBR had taken up the ambitious project of construction of at least 600 miles of railways. It was given a concession to construct lines from Kolkata to the river Ganga at Kushtia and ultimately to Dhaka, together with a branch to Jessore. At the same time the company was empowered under an act of incorporation to increase its capital to £6 million. So important was this EBR project to the government that it was submitted before the home government in the utmost possible detail so that there was hardly any room for dispute between the company executives and the government officers.[9] The enthusiasm for railways continued unabated until the twentieth century. On the other hand, efforts to improve the waterways were relatively feeble – as reflected in the fact that whereas total expenditure on the improvement of navigation facilities was around £5 million during the last two to three decades of the nineteenth century, expenditure on the railways during the same period had exceeded £200 million.[10] The EBR was treated relatively favourably and in 1928 had received about two-thirds of the total sum asked for in the year 1925–30, while other lines had received less than half their demands.[11] Though conditions in Bengal were 'more favourable for the improvement and extension of such navigation facilities than in any country in the world', the bias towards railways in East Bengal continued. When the construction of the Eastern Bengal Railway was projected, the computations concerning the amount of tonnage it was likely to carry were based on the returns of the Eastern Canals.[12] It was calculated that more than a million tons of produce were transported annually to the port of Kolkata from the East Bengal districts and that at least 40,000 tons of imports were distributed over the same territory as return cargo.[13] However, it was the EBR, not the Eastern Canals, which began to receive patronage. The relationship between the railways and the ever-changing deltaic landscape can be fruitfully examined keeping the above context in mind.

The water bodies and the railways: the range of contestations

First and foremost among challenges to the emerging railway system in Britain was that from those who saw waterways, especially canals, as ideal routes for commodity transport. As the proponents of railways listed the potential advantages of the establishment of railways, so their opponents pointed to what the railways could not possibly deliver. For example, the *Edinburgh Review* in October 1834 published a set of arguments in favour of the railways in which the author identified fourteen sectors of the national economy where he thought the railways could play a pivotal role. In the following year, all these arguments were refuted by one R. Cort, who seemed to be convinced by the enormous advantages of continuing with traditional waterways.[14] Cort even asked whether there was 'nothing wrong in spreading a false system of conveyance in every quarter of the globe, as well as deserting the ancient thoroughfares of the kingdom?' Central to the conflicts between the protagonists of railways and canals was the issue of capital investment. Cort revealed that the amount of capital invested in the internal navigation of Britain was no less than £30 million and he demanded that in the unavoidable case of expansion of the railways the canal interests should be protected by parliamentary legislation.[15] Since Parliament itself had safeguarded the investments in the development and maintenance of canals, it was under compulsion to protect the canal interests. On the other hand, the wonder and prospects of new steam technology were drawing fresh attention from numerous private investors. Parliament's dilemma was evidenced in the Railway Consolidation Act of 1845. The act proposed that railway companies could alter or divert the course of any rivers, brooks, streams or water courses and any branch of a river for the purpose of constructing tunnels, bridges and other passages as the companies thought proper. However, the act divided the different water courses into 'navigable' and 'non-navigable' sectors and the railway companies were given liberty to deal with non-navigable water bodies only. The navigable water bodies did not come under the jurisdiction of the railway companies.[16] The water regime of Britain was thus conveniently shared by both the railway and canal interests.

That is to say that the capitalist interests representing the canals and the railways were apparently given a fair share of security of investment, but in the process the ecological regime of water as a whole was undermined, since 'navigability', rather than the intrinsic ecological value of water, provided the context for a solution. The legal implications of the tensions between the two major modes of transport in Britain had a bearing on the way the railways and waterways were seen in relation to each other in Bengal. Yet, the impact of the railways on the water regime of Britain was trifling in comparison to that of the Bengal Delta; on the one hand, because of the nature of the landscape and fluvial conditions of Bengal, and on the other, as a result of political and economic conditions. In Bengal water bodies of different length and width were so lavishly spread across the delta that there was no real question of private investment in maintaining canals and similar waterways. Therefore, when the railways were being spread across the delta, they faced no great opposition from private investors or other financial interests as had been the case in Britain. The absence of a rival capitalist interest in Bengal left the railway establishment as the sole arbiter of its own operation. In the context of this monopoly of the railways, which drew obsessive approval from both metropolitan and colonial administrations, the Bengal Delta saw the railways interfering not only with 'non-navigable' water bodies, but also with 'navigable' water bodies. Within a decade of the coming of the railways to the delta, the Bengal government legislated that it was lawful for the Lieutenant-Governor of Bengal to order the blocking of any navigable channel.[17] Canal officers were enabled to close any water courses for 'public purposes'. Diversion of rivers was also stipulated in connection with the building of the railways.[18]

The influence of capitalist enterprises with respect to railway construction was evidenced in the way the railway engineers linked the water regime of the region to professional knowledge. The position of the engineers can be examined by looking at the way they negotiated the twin realities of safeguarding financial investment and constructing the railways in a very fluid landscape. Some of these issues were raised in the lectures given by senior engineers to the students at Sibpur Engineering College. In one series of lectures, S. Finney, a manager of the EBR, noted that while laying out a railway line in some parts of India was a troublesome operation, in Lower Bengal it was an extremely simple matter, so easy in fact that the work could be done 'without proper care' – a surprising statement since the many rivers and waterways should have posed considerable difficulties to the construction of the railways.[19] What considerations then, led the engineers to think that railway construction in the Bengal Delta was such a simple matter? Apparently, an engineer's preference in selecting a site for constructing a railway line was subject to modifications based on the needs of trade and administration. Finney told his students at the Sibpur Engineering College that the choice of route would depend, first, upon traffic prospects, and second, on economy in construction and operation. He reminded them that the best engineering was not that which made the most splendid or even the most perfect work but that which made a work that answered 'the purpose well at the least cost'.[20] Another engineer, H.W. Joyce, suggested that the railway was 'purely a business investment constructed to pay dividend, and to satisfy them in regard to the likelihood of a projected line to furnish these dividends or not is one of the duties of the Engineer'.[21]

As engineering sophistication appeared secondary to investment portfolios and commercial schemes, and as the water regime of the delta lay within the jurisdiction of the railway companies, the simplest solution to the extension of the railways seemed to be a massive extension of the embankments on which the railways were to run. Finney reminded his students that practically all railways in Lower Bengal were built on embankments.[22] The 3500 miles of railways that would traverse the Ganga-Brahmaputra watershed represented a similar length of embankments. But how easy was it to erect and manage these embankments on Bengal's deltaic landscape with rivers draining in a myriad of directions? To what extent, for instance, could the waterways and embankments coexist or not? How did the engineers justify their claims that railway construction in the plains of the delta was easier than in other places and that embankments appeared to solve many of the problems associated with railway construction?

One of the issues that vexed the engineers was how much space should be provided through embankments for the free flow of water. Finney thought that the number of waterways to be provided must depend on the importance of the railway traffic. If the line in question was of 'first class importance' and the traffic was heavy, then an interruption of communication was calamitous, and waterways must be broad, since the flood could breach the embankment and the loss from an interruption could be enormous. If, on the other hand, a branch line was being constructed where the traffic was small, 'some risk of an occasional interruption' was 'justifiable'. Another related question of whether the waterways should provide for ordinary floods or for abnormal ones occurring at long intervals also depended on the importance of the traffic that was anticipated for the railways, Finney suggested.[23] These questions of providing waterways through the railway embankments took a philosophical turn when Joyce explained the merit of a 'rational' instead of 'empirical' method in determining the width of culverts for passing water. He criticized the prevailing methods of 'offhand judgment' by engineers in mapping the adjacent water-bodies before ascertaining the size of the culverts. Joyce thought that these 'hopelessly unscientific and extravagant' methodologies could easily lead to the construction of unnecessarily large culverts.[24] In his 'rational' scheme of measuring the required space for waterways in small culverts, he considered the following issues: the area of the basin the culvert must drain; the maximum rainfall to be expected in the locality and the time of its continuance; the velocity of the flow; and the portion of the water falling on the drainage area that would reach the culvert, commonly called 'run-off'. Noting that varying conditions of soil, surface and seasons caused the run-off to vary in different years and at different times, Joyce suggested that the notion that maximum run-off must be provided for was a simplification. Considering the case of small areas with a surface rendered impervious by considerable previous rainfall and other causes, Joyce assumed that 90 per cent of the total rainfall would pass through the culvert. But in a larger area, where the extreme distance between the watershed and the culvert might be as much as three miles, not more than 60 per cent of the rainfall would reach the culvert. Therefore, Joyce thought that the empirical formula commonly used to determine the size of sewers did not seem applicable to determining the run-off in the case of culverts, as the conditions in the two cases were quite different. After thus ascertaining the maximum quantity of water to be accommodated, Joyce went on to determine the size of the culvert by 'well-known methods, modified by conditions encountered at each location', and by such a factor of safety as the engineer thought prudent. The priority of the 'safety' of the railways over ecological concerns was made particularly clear by the following statement of Joyce:

> There must be some limit where the expenditure of money to provide for remote contingencies becomes unwise. No one doubts the possibility of earthquakes in any locality, but in most places the contingency is so remote that we do not consider it advisable to build our houses to resist earthquake shocks. The wisdom of designing railroad culverts of such size that they will pass floods that are only likely to occur in each 25 or 30 years may well be questioned.[25]

With such a stress on rationalization within the railway establishment, it was not surprising that in 1938, for example, at Dhulia in Murshidabad the authorities resisted the appeal by local people to cut the embankment to enable stagnant water to drain out through the railway line. The local rail authorities insisted on 'modern' knowledge and noted that whatever the old people of the locality might believe, more culverts could not be the solution.[26] On another occasion, a government executive engineer remarked that the local inhabitants were neither accurate nor keen observers. He remarked that the only trained observers were railway officials but their evidence could not be accepted as conclusive because their main object had been 'to a defence of the railway'.[27]

Conflicts between 'rational' and 'traditional' knowledge of the deltaic landscape were further highlighted by the apprehension that excessive waterways around the railways posed a threat to the lines. In the early 1920s, the government of India asked the government of Bengal about the influence of railway construction on public health and advised the appointment of a committee to consider the sanitary conditions of any line to be constructed. This led the government of Bengal to appoint ten surveyors under the direction of the Chief Engineer of the Public Health Department and with the support of the Railway Board. But before sanctioning the survey and the appointment of the surveyors, the Irrigation Secretary opposed the proposal on the grounds that the only way to drain accumulated water in the borrow pits was to join them by water channels to the nearest rivers, and that when these channels were deep, they would pose a danger to the lines. The file was then sent to the Department of Public Health which ultimately failed to press the authority to carry out the project on health grounds. Meanwhile, the prospective engineers were reminded by Finney to avoid cutting deep pits as far as possible and that under no circumstances were they to be continuous. Finney told his students that a substantial buffer 'must remain between every pair of pits, or you will cause great risk and danger later on by a flow of water parallel to your embankment'.[28] The idea of 'saving' a railway line from the rush of water remained the major priority for colonial railway engineers. It was, therefore, not surprising that minor rivers or water courses were often blocked altogether during the construction of the railways.[29]

As early as the 1920s Bentley pointed out that the 'blind' way of building roads and railway embankments without adequate culverts had resulted in the division of the country into 'innumerable compartments' and that it was extremely difficult for rainwater to flow from one compartment to another. Every year, Bentley added, the floods increased in severity and he warned that unless remedial measures were adopted, the region would cease to exist as the richest rice producing area in the country.[30] About the same time the report of the Royal Commission on Agriculture in India observed that embankments led to deterioration in rivers by raising their beds and advised the government to set up a Provincial Waterway Board for Bengal to take care of these problems.

The board came into being in 1932 but, as a contemporary critic observed, it represented only commercial interests and no provision was made for the representation of the Public Health Department, which had been keenly observing the adverse impact of embankments.

Case study I: the Eastern Bengal Railway (EBR)

The first railway line in eastern Bengal, running from Kolkata (Sealdah) to Goalundo, was constructed on the floodplain of the Ganga, which, along with its numerous branches and tributaries, flowed to the Bay of Bengal. Initially there were almost no outlets for the passing of water through the embankment on which this line was constructed. The necessity of outlets through the embankment was felt every time there was a flood. About 2000 lineal feet of 'opening' were added between 1868 and 1885. After the floods of 1890, a further 400 feet were added.[31] The north-western segment of the EBR contributed to the deterioration of the water regime of northern Bengal through the impact of the railway on the Chalan beel.[32] The Chalan beel was a vast deep hollow with a watershed of about 1547 square miles, lying in the districts of Rajshahi and Pabna, where a very large portion of the drainage from about 47 rivers of northern Bengal converged.[33] Besides being a giant junction of numerous waterways, the beel also served as a springboard from which many rivers flowed further south and east to meet finally with Padma or Brahmaputra. With the waterways that converged in it from the north and north-west and with those that exited from it towards the east and south-east the Chalan beel formed a water regime that reserved and cleared the drainage of almost half of the Bengal Delta.[34] By the beginning of the twentieth century the beel was surrounded, with the EBR main line to the west and the Santahar-Bogra line in the north. Since the beel filled from the north-west and south-west, this process was intersected by the Bogra-Santahar branch line and the EBR main line, and since it drained in a south-easterly direction to the Brahmaputra, the drainage was intersected by the Sara-Sirajganj branch line of the EBR.[35] The natural drainage of this part of the delta met with formidable obstacles in the form of the embankments that were necessary for the construction of the railways in low-lying areas. The situation was further aggravated by the reduction of the number of spans on the bridges of the EBR following the construction of a broad gauge line between Atrai and Santahar. In this area the total existing outlets in the early 1920s were reported to be 440 feet as compared with 967 when the line was first constructed.[36] In the case of the southern branches of the EBR, for instance, although the combined catchment area of the waterways was 1.5 square miles between the Dadshi and Pachuria railway stations in Khulna, there were only four openings. Although government officials considered the openings 'adequate', the actual measurement of the four openings revealed that there were two pipe culverts each of 1.6 feet diameter and two girder bridges of $1 \times 12 = 12.0$ feet and $1 \times 20 = 20.0$ feet diameter for the entire catchment area.[37] Such inadequate openings were found in almost every culvert or railway bridge across the delta.[38]

The main EBR line (Kolkata-Siliguri) ran across the Rajshahi plains, the natural slope of which was towards the east. Railway embankments interrupted the natural drainage of the country, blocking a great volume of water, the effect of which was the complete destruction of aman (autumn) paddy, the only subsistence crop of the land. The water took a long time to escape through the culverts and bridges which were very few in number. Its passage to the river Brahmaputra was further obstructed by the Sara-Sirajganj embankment and consequently destroyed the crops of another area. A part of the line also passed close to the Chalan beel and thus prevented its flush water from draining to the Brahmaputra; the result of this interruption was the speedy silting up of the beel and the consequent reduction of its water-holding capacity. This reduction, according to a government official, was one of the factors underlying the frequent flooding following the construction of the Sara-Sirajganj railway. The cultivators of the vast area to the north-west of the Sara-Sirajganj line were forced to stop the cultivation of rabi (winter) crops as the fields did not dry up in time for cultivation. They resorted to sowing seeds in the mud. As one official reported, 'Any one with an ounce of knowledge in Agriculture could easily imagine how a crop of mustard, lentil, wheat or barley fares if thus sown.' He made clear that he was not speaking of years of abnormal rainfall, but of normal years.[39]

In 1928, a devastating flood took place in Bogra and Rajshahi, beginning with excessive rainfall on 21 August in Bogra. The most affected area lay on both sides of the EBR between the Hilli and Nator rail stations. The water east of the EBR built up behind the embankment of the Bogra line. Meanwhile, heavy rain fell in Rajshahi on 24 August, which added to the flood water draining from the Bogra and Dinajpur districts in the upstream and caused all of northern Bengal to be flooded. Here too the railway embankment prevented the flood-water from draining away. Between 1300 and 1400 square miles of land were affected and more than 200 square miles of crops were destroyed.[40] The Bogra line ran almost directly east, blocking the natural flow of water of a part of the country which sloped from north to south. The resultant waterlogging damaged the area's only crop (aman rice). But as the flood-water slowly moved further down – either through an insufficient number of culverts and bridges or by overtopping or breaching the embankment – it was again obstructed by a part of the Sara-Sirajganj railway embankment. The same degree of devastation of crops occurred in these tracts until the water finally found its way into the Brahmaputra (see Figure 6.3 above).[41]

Case study II: the Assam-Bengal Railway (ABR)

While the EBR ran mainly across the Ganga watersheds, the ABR covered the eastern half of the delta, which was under the sway of the Brahmaputra and the joint flow of the Meghna and the Padma (Ganga). The ABR entered the plain of the delta through Mymensingh and finally reached Chittagong, connecting parts of the Dhaka, Comilla and Noakhali districts. The ABR line from Bhairab Bazar to Kishorganj crossed the spill of the Meghna over its right bank. The line gradually diverted towards the west and ran between the old Brahmaputra and the Ghorautra rivers. The Ghorautra was a live effluent river of the Meghna and it carried almost the entire run-off of the southern slope of the Garo Hills. There were a few beels, which drained into the Ghorautra river through the openings in the embankment. It appeared that due to the obstruction in the spill of the Meghna, the spill of the Ghorautra river spread over the area and was obstructed by the railway.[42]

Early and high flooding reportedly became an annual feature of the vast surrounding areas following the construction of the Mymensingh-Bhairab Bazar line. Untimely flooding damaged young paddy as well as jute.[43] According to the executive engineer of the Mymensingh division the drainage problem would be solved if the spans of the two railway bridges along the main line (Chittagong-Akhaura railway) were increased.[44] During 1931–32, on account of a heavy rainfall in the Assam Hills the areas on both sides of the Brahmaputra up to a distance of 20 miles from the river were submerged and about 35 per cent of the standing crops of paddy and jute were damaged. About 1300 square miles and a population of over 500,000 were affected. All the subdivisions and almost all the thanas of Tangail suffered the loss or damage of standing crops.[45]

The situation in Comilla was more critical. The district was bounded on the east by the Tripura (Tippera) Hills and the fringe of lands on the west formed a gradual slope which had the Titas and Gumti rivers as its outfall. The ABR main line ran almost parallel to the Tripura Hills and cut through the extreme eastern boundary of the subdivision of

Brahmanbaria where this portion of the railway formed a ridge on the eastern slope, thus obstructing the free passage of the surface water of the slopes which fell into the Titas and the Gumti rivers and other drainage canals lying to the west of the railway. A sub-divisional officer reported:

> The one physical fact that stands out in differentiating the present from the past is the branch railway line from Akkaura [Akhaura] through the Brahmanbaria to Ashuganj which passes through as many as six union boards. From 1910 (since the start of this line) there has been an appreciable deterioration in the agricultural condition and the health of the people in general. Malaria fever and kala-azar are found to be in prevalence in the villages situated alongside the railway lines. In normal years the difference in the water levels on the two sides of the branch and main lines is more than one cubit.[46]

The magnitude of the problem in the district of Comilla in general and Brahmanbaria subdivision in particular could be appreciated from the fact that these areas had to cope with a huge amount of surface water from numerous sources as well as the rainfall in the district itself. Water gathered here from world-record rainfall in Cherapunjee and Silchar, rainfall in Sylhet and Cachar, and rainfall in the Tripura Hills. Along with these sources came floods in the river Gumti and even the surface water of the tract on the border of the Meghna river. With a new railway line in place the accumulated water drained too slowly, resulting in the delay of sowing and transplantation of paddy. Waterlogging led to food shortages and distress in 1915 owing to the failure of crops. In 1926, the low lands between the two railway lines were completely under high water for an unusually long time and the standing crops, mainly jute, were destroyed. The situation was further aggravated by an insufficient number of bridges on the ABR.[47]

The scenario in Noakhali and Chittagong, the south-eastern segment of the delta, was no different. It was reported, for instance, that the culvert just south of Baraiyadhala railway station was too narrow to allow the water of the Chittagong Hill streams to pass freely. The obstruction caused silt and gravel to deposit on the bed of the channel which was six to eight cubits deep. The water, therefore, ran over the neighbouring fields. As a result betel vine cultivation was damaged 'year after year' on account of fungal diseases that resulted from the submergence of the plants. The loss in the year 1922 amounted to several million rupees and entirely deprived the Barais (the local traders in betel) of their livelihood. In Noakhali, owing to the narrowness of the railway bridges, drainage was obstructed and thereby caused damage to the crops in the fields of at least seven villages. In 1915, some 20 square miles in the Feni subdivision were flooded with the result that the aush paddy was damaged and this, along with other factors, contributed to a food shortage and starvation in the region.[48]

Case study 3: the Hardinge Bridge

The problem with railway embankments was not limited to inadequate openings through culverts. Larger bridges also posed a considerable menace to the water system of the region. As Bengal rivers carry huge amounts of silt, deposition can be dependent on quite trivial factors, such as a fallen tree or a sunken boat. The pillars and piers of railway bridges had a remarkable effect on the process of siltation.[49] Thus the activities relating to the control of rivers on behalf of the railway establishments contributed to the recurrent problem of river-bank erosion. The stretch of continuous revetment at Pabna which had to bear the full force of the main stream did not prevent the river from cutting away large chunks of the bank immediately below. The great Goalundo spur, constructed in the 1920s at a cost of £120,000, did not save Goalundo; it rather caused a scour to the depth of 180 feet, was itself destroyed and failed to save the railway settlement it was meant to protect.[50] The Hardinge Bridge over the Ganga at Sara (on Sara-Sirajganj line) provided an ideal example of the problems associated with railway bridges. Well before the opening of the bridge in 1915, the fishermen of the Ganga feared that it would slow the current and lead to the deposition of silt and drying up of beels and inland fisheries.[51] How well-founded were these fears?

The Hardinge Bridge project was formulated by the turn of the nineteenth century, but the actual construction started only in 1911 (Figure 6.4). The bridge was expected to cater to the need of connecting the most important jute-growing areas of the delta with Kolkata while avoiding the double trans-shipment and delay on the Ganga. The line was also expected to lead to a large increase in passenger traffic between Darjeeling and Shillong and Kolkata. The bridge was considered the 'most important engineering scheme' in India and in some respects 'one of the most notable in any part of the world' at that time. The difficulty that confronted the engineers during the construction did not relate to the question of how to span more than a mile of Ganga water but to the question of how to control the river which frequently changed its course.

Figure 6.4 Photo of a construction site of the Hardinge Bridge over the Ganges
Source: Centre of South Asian Studies, Cambridge. Alexander Collections.

Therefore, the first thing the engineers attempted was to harness the strong current of the Ganga. Since the annual rise of the river in flood time was about 31 feet and a maximum flood discharge at the site of the bridge was 2.5 million cubic feet per second, this was a gigantic project. A pair of guide banks were constructed at the bridge site to prevent further lateral movement of the river, and a revetment of the banks was built at the two ends of the projected bridge (one at Sara Ghat Station and another at Raita Ghat station). It was calculated that the amount of stone used in 'pitching' these guide banks of sand and clay would fill a broad-gauge train extending from Kolkata to Darjeeling, a distance of around 300 miles. One of the noticeable features of the project was the approach work, which alone cost 8.4 million rupees. On the left bank the approach was about 4 miles long and for 2000 feet of that length the approach had the unusual height of 50 feet above the surrounding area. On the right bank the approach was three miles long with similar characteristics. The project employed as many as 25,000 coolies at a time. The bridge stood on 16 piers, 63 feet long and 37 feet wide, for which well foundations were dug 150 feet deep. The great depth of the wells, the deepest in the world at that time, was necessary in the context of the enormous scouring of the river. The piers were formed out of concrete blocks above the steel caissons, and steel trestles above the high flood level. The wells required 36 million cubic feet of 1.50 stone ballast, 2 million cubic feet of sand, 125,000 casks of cement and 7906 tons of steelwork.[52]

The modern bridge-building technology and engineering as exemplified by the Hardinge Bridge, however, came at a cost. For hundreds of square miles around the bridge the land suffered in various ways. It was reported that the engineering works caused a 'complete upset in the Ganges below and above'. In July 1928, at Lalgola of Rajshahi, for instance, the Ganga cut away a strip of land 100 feet wide and 55 feet deep in one night: 'this erosion extended for a good many miles above, where the deep channel ran along the bank. Many spurs in this province were abandoned because of the damage they caused'.[53] Surprisingly, the effect of the bridge was felt as far north as Murshidabad. It was reported from Murshidabad, where a lot of aush paddy perished in a devastating flood in 1939, that the direct cause of the flood was the overflow of river water from Padma and its tributaries, that is Jalangi, and that the overflow was partly due to the 'heavy back-rush' of currents in the river caused by the obstruction at the Hardinge Bridge.[54] The damage to crops caused by the flood in Murshidabad was immense: 90 per cent of aush rice, 25 per cent of aman and 75 per cent of jute were lost.[55] In 1941, cultivators complained that flood-water had 'banked up' against

Figure 6.5   Hardinge Bridge, early twenty-first century

the piers of the bridge and inundated the fields for miles around.[56] More alarmingly, the gigantic construction work had an adverse impact on the water regime of the region in general. For instance, following the construction of the bridge, a few channels of the Dhaleswari river had unexpectedly flowed into the neighbouring beel. The strong current of water along these channels regularly caused extensive damage to paddy crops cultivated in an area of about 20 to 30 square miles. At the same time, the bridge enormously contributed to the deterioration of the Ganges (Padma) by encouraging the deposition of silt. Over the years, siltation has occurred to such an extent that motor cars can now be seen plying under the bridge in the dry seasons (Figure 6.5).

* * *

As we have seen, the results of the ecologically unsustainable development of the railways in the fluvial landscape of the Bengal Delta were disastrous for agrarian Bengal. The agricultural decline in Bengal from the early twentieth century should not be attributed to the railways alone, but there are indications that by contributing to the deterioration, drying up and death of various types of water bodies, the railway embankments may well have indirectly contributed to the overall agrarian decline in Bengal. Cultivable waste land in the Bengal

Delta increased over the first few decades of the twentieth century, for instance.[57] In the case of western and central Bengal, an explanation of this might be that these regions lost population due to the moribund state of the deltaic landscape and the spread of malaria. But historians of Bengal have not explained why cultivable wasteland existed in eastern Bengal, where the rate of population growth was significantly more than in other areas and where land was still comparatively fertile. Deterioration in the water system led to flooding, crop failure, waterlogging, and loss of navigation and marketing facilities. Whereas eastern Bengal exported rice throughout the nineteenth century, by the beginning of the twentieth century it had to import rice for subsistence. Historians have argued that one of the reasons for the Bengal famine in 1943 was the stoppage of the importation of rice from Burma during the Japanese occupation. But the unasked question remains why, in the first instance, had rice production so declined in the region that it was dependent on such imports. It might be assumed that decline in the ecological regime led to decline in both commercial and subsistence produce which, in combination with other factors, ultimately led to the famine. One needs to consider the role and potential responsibilities of the railways, with their long arms of embankments.

The railways in India in general and in Bengal in particular expanded amidst overwhelming approval from investors in London and colonial officials and an emerging middle class in Kolkata.[58] A positive perception of the railways was so wedded to the state-formation process that there was hardly any room for critical voices. An overt emphasis on the potential of the railway as a system and symbol of modernization met with sharp reaction from those who were sceptical about modernity itself. But the emerging colonial knowledge of the railways – among advocates and opponents alike – failed to grasp the extent of the threat of the railways to the ecology of the region. Karl Marx hoped for an extensive irrigation system, emerging as a result of digging the soil for railway embankments, but he did not foresee the problem of drainage. On the other hand, Gandhi could well perceive the railways as carriers of fatal diseases, but he was not in a position to appreciate the capacity of the railways to create disease by destabilizing the fluvial ecology.

Surprisingly, a scientific approach to the problem of embankment in the deltaic landscape of Bengal came from a segment of colonial officialdom. In 1846, a committee appointed to examine the problem of embankments in Bengal put forward some strong arguments against any barrier in the flat and fluid landscape. Consisting of two engineers and a botanist, the committee proposed a 'return to that state of nature, which, in their opinion, ought never have been departed from'. To achieve this goal, the committee recommended the total removal of all existing flood embankments to allow the free flow of water. The proposed system, to be built in consonance with 'local experience', was tantamount to reversing the existing system of embankments by substituting them with drainage.[59] However, this soon gave way to a solution favouring the construction of the railways on high embankments – a clear reflection of the enthusiasm for modern technology and the interests of powerful investors. When C.A. Bentley reiterated his concern in the early twentieth century over the negative impact of the railway systems on the water regime of Bengal and on its ecology and agriculture, it was too late, for the railways had already become an integral part of public life.

7
Fighting with a Weed: the Water Hyacinth, the State and the Public Square

If the railway embankments led to the deterioration of the water system of the Bengal Delta, an Amazonian weed, the water hyacinth (*Eichhornia crassipes*), further complicated the ecology of the region. But the story of the water hyacinth, which was present across four continents in the early twentieth century, is also interesting in the insights that it provides into the way in which the colonial state in India dealt with a biologically alien waterweed.[1] Such a study is necessary in the broader field of environmental history because, following Alfred Crosby's seminal work on biological exchange, a lot more focus has been put on the relationship between plant transfer and imperial expansion than on the actual encounter between a secure colonial state and an invasive plant.

This chapter examines the ambivalent position of the state regarding the destruction or scientific utilization of the water hyacinth; the predicament of government in its quest for legislation to contain the weed; and the complications and failures of legislative attempts at eradication. In examining these issues, the chapter focuses on the ways in which different bureaucracies and different realms of science as well as private commercial interests imagined, constructed and represented the problem of species invasion in a colonial context. In such a context, a wealth of competing players were at work, contradicting one another, struggling over bureaucratic power and funding, and attempting to further and extend their administrative reach. The hyacinth was caught up in these machinations in interesting ways, though it was never tamed by the state.

The growth of the water hyacinth in the Bengal Delta

The water hyacinth may have been introduced into East Bengal by George Morgan, a Scottish migrant and jute merchant of Narayanganj,

140

an industrial district in Dhaka, around the turn of the twentieth century. Morgan was impressed by the beauty of the flowers and leaves of the plant and brought it with him on his return from Australia.[2] Another narrative has it that the hyacinth was brought to Calcutta Botanic Garden from Brazil in the 1890s and that, at a later date, some ladies, being attracted by its flower, collected and transplanted these weeds to their gardens in Dhaka.[3] Some believe that the weed made its way downstream to the delta through the river Brahmaputra from Assam.[4] The rapid spread of this weed in Bengal at the outset of the First World War has also been attributed to the Germans, who wanted to weaken the British by 'killing their Indian subjects', hence it became known as the *German pana* or German weed.[5] As implied later in the chapter, a transnational company might also have introduced the plant.[6]

In 1914, the Narayanganj Chamber of Commerce considered the menace of the weed as one of 'sufficient importance' to bring it to the government's attention. By 1920 it was acknowledged by both government and non-government agencies that the water hyacinth had been 'choking up the natural arteries of trade, impeding agricultural operations and menacing the health of the people' in most parts of East Bengal.[7] In the 1920s, while a Bengali journalist compared the effects of the weed with malaria epidemics, which were a major cause of mortality in contemporary Bengal, a colonial official considered it the most pressing problem after the anti-colonial terrorist movement.[8] A conservative estimate revealed that in 1936 the hyacinth covered an area of over 4000 square miles.[9] The weed was mostly prevalent in the active delta, which comprised an area of around 35,000 square miles – implying that the hyacinth covered about one-ninth of the total deltaic plain. If the lands covered by homesteads, office buildings, temples and mosques are excluded and only water-bodies and agricultural lands adjacent to them are considered, the coverage would have been proportionately higher.[10] (See Figure 7.1.)

As the spread of the water hyacinth was left largely unchallenged, the devastation it caused to crops and cultivation processes remained unchecked. In the district of Mymensingh, for example, cultivators gave up producing any crop over an area of 100 square miles because of the extensive damage caused by the water hyacinth 'year by year'. In Khulna beel (marshy low land) areas, paddy cultivation was rendered difficult, and low-lying paddy suffered damage from the encroachment of the plant.[11] The people of Nasirnagar in Comilla district petitioned the government, alleging that since as early as 1915, crops had been destroyed over a very large tract of their land by flooding and the

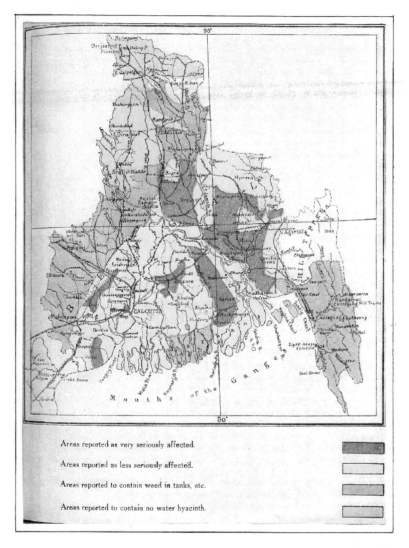

Figure 7.1 Map of Bengal showing areas affected by the water hyacinth, 1920s
Source: Kenneth McLean, 'Water Hyacinth', *Agricultural Journal of India*, XVII (1922).

water hyacinth.[12] A large quantity of paddy grown in the Arial beel of Munshiganj of Dhaka District was reported to have been destroyed by the weed.[13] The hyacinth from the Kumar river destroyed paddy and jute plants across an area of more than 174 square miles each year.

It was also alleged that inland navigation and the cultivation of jute and paddy of the aman variety had become difficult due to the pervasive presence of the water hyacinth. It was reported in 1926 that 15 to 20 per cent of the aman paddy was being damaged 'year after year'.[14] The mover of the Bengal Water Hyacinth Bill (1933) noted that 'some time ago' the annual damage done by the water hyacinth in Bengal was estimated at about six crore rupees (1 crore = 10 million) and at the time of his speaking it was 'very much more'.[15] This was not an exaggeration, since the water hyacinth was particularly destructive of beel paddy that grew in abundance in the delta.[16] In a region that mostly comprised deltaic low lands, being uniquely fit for a range of rice species, the chronic presence of the water hyacinth contributed to what has recently been termed as an 'economic depression'.[17] (Figures 7.2, 7.3, 7.4 and 7.5.)

The problem with the weed was further complicated because of an insufficient flow of water in the region. Where embankments, both protective and railway, were erected with few water outlets – which was usually the case – currents of water were blocked or reduced. In places where canals or smaller streams were blocked by the pillars and plates of locks and sluice gates, siltation took place, providing a congenial home for

Figure 7.2   Photo of a boat loaded with jute navigating through the water hyacinth
Source: A Short Survey of the Work Achievements and Needs to the Bengal Agriculture Department, 1906–1936 (Calcutta, n.d).

Figure 7.3 A peasant struggling to ply a boat through the water hyacinth
Source: Kenneth McLean, 'Water Hyacinth', *Agricultural Journal of India*, XVII (1922).

Figure 7.4 Photo of a khal choked with water hyacinth
Source: Kenneth McLean, 'Water Hyacinth', *Agricultural Journal of India*, XVII (1922).

Figure 7.5 Photo of a tank in Comilla town choked with water hyacinth
Source: Photo taken by the author in autumn 2002.

the water hyacinth to take root and to multiply. Ditches alongside railways and roads under district authorities were also thought to be places of 'infection'.[18] During the months of Falgun and Chaitra (roughly in spring) cultivators used to destroy all the hyacinths that grew or accumulated on their land; but the hyacinths on the khas (private) lands of the landlords and of the government remained intact. With the arrival of the rainy season, the weed 'grew far and wide and destroyed the crops of the poor cultivators'.[19] In 1946, it was estimated that the hyacinth was destroying crops and fish worth at least 10 million rupees every year.[20]

In the field of public health, the water hyacinth was considered responsible for spreading influenza and various water-related diseases.[21] However, in response to the suggestion that the hyacinth contributed to the spread of cholera, C.A. Bentley, the Sanitary Commissioner of Bengal, thought that it could only have had such an impact if its presence encouraged the pollution of water with human excrement, which he doubted. Bentley thought that the only possible indirect way in which the plant could cause cholera would be by shading polluted water from sunlight, thereby interfering with the natural process of purification, which took place in a few days in the case of water exposed to sunlight and air. But Bentley thought this to be 'purely hypothetical' and though he admitted that the weed was a 'great nuisance', which needed to be dealt with, he failed to acknowledge its link to cholera.[22] As far as the relationship between the water hyacinth and malaria was concerned, Bentley noted that water thickly covered with hyacinth rarely showed any evidence of the presence of any anopheles mosquito larvae.[23] However, a report by S.N. Sur, a field-level public health official in the Malaria Research Unit in Bengal, contradicted Bentley's assumptions. He observed that the prevailing malarial condition was mainly due to the stagnation of water hosting the water hyacinth, which favoured the growth of mosquito larvae by 'reducing the temperature of the water as well as giving shelter against their natural enemies'.[24]

In addition to its considerable impact on health, the water hyacinth also affected public nutrition through its impact on fish stocks. By thriving in the pukurs (tanks or ponds) of the countryside during the rainy seasons, it not only polluted drinking water but also posed a danger to fish populations. Its prevalence was considered to be one of the reasons underlying the rapid fall in fish production in Bengal.[25] Along with human health, the health of cattle, which were the backbone of agriculture in Bengal, appeared to have been affected by their eating the water hyacinth. J. Donovan, a district magistrate in Barisal, noted that he had never seen more miserable cattle than those of East Bengal. He learnt from the veterinary officer of the district that in the absence of adequate grazing, the cows were suffering from indigestion as a result of eating the water hyacinth.[26] The link between the water hyacinth and decline in agricultural production and health was described by a local witness in these words:

> The inroads of savage army, through the frontiers, the incursions of a Timurlane, carrying fire and sword into the country, were nothing compared to the inroads of those tiny plants, floating down the East Bengal rivers ... creeks, canals and small rivulets had been clogged and choking up ... even costly careful clearance, twice a year, was not able to arrest its growth ... during flood tides, these plants get into fields and within a few days, by first multiplication, cover them entirely to the destruction of rice and other crops rooted on the earth ... Eastern Bengal, the granary of the Province and hitherto the healthiest portion of it, is being rendered desolate by the bringing of malaria by this plant...[27]

The official perception of the speedy growth of the water hyacinth was that deltaic East Bengal provided an ideal physical environment.

In an attempt to examine the capacity of the weed to grow in different environments, in 1920 its seeds were tested in a government laboratory for germination on dampened blotting paper, in water, in mud and in damp soil. The seeds were kept under observation for one month during February and the tests were made under both ordinary atmospheric conditions and in the incubator at a temperature of 86° F. The hyacinth germinated in 'all conditions' and it appeared to be 'perfectly formed and healthy'. As the weed was able to germinate in different environments, so was it able to spread itself by virtue of its bladder-like leaf stalk and sail-like leaves; the former enabled it to float and the latter, with the help of wind, enabled it to travel into new areas. An observation team made up of officials and local people found in their experiment in the Turag river in Dhaka that the weed could travel at the rate of three miles per hour. Apparently, a single root of the hyacinth could cover an area of more than 600 square yards in the space of a few months. It was observed in a government report that if there were any case of death of the water hyacinth, it was due to its being 'overgrown and submerged by its progeny'. Nothing except severe frost could weaken and destroy the weed, and frost was exactly what was wanting in this tropical delta.[28] Such an official representation of the 'extraordinary biological strength' of the water hyacinth did not come as a surprise, since this was one of the ways available to cloak the government's vulnerability in containing the weed effectively.

Though the above discussion indicates the range of potential problems to which the East Bengal environment was exposed because of the water hyacinth, it remains difficult to estimate the extent of its actual contribution to the decline in agriculture. Officially the weed was perceived either as a mere nuisance or as a potentially profitable plant; it was not seen as a contributory factor to declining agrarian production. This is probably why no comprehensive effort was made to monitor the statistics of the growth and impact of the weed. It is easy to investigate, for instance, rice or jute production from well-preserved government statistics, but it is not so easy to gain an accurate picture of the impact of the water hyacinth, although it had already become a public issue in the 1910s. It is, however, possible to obtain information regarding various government efforts to combat the weed and relevant responses from the wider public arena. We will now turn to these issues.

Eradication or utilization?

From the beginning of its fight against the water hyacinth, the government of Bengal had to cope with the dilemma of whether the weed should be completely eradicated or be fruitfully utilized. The first working proposal towards utilization came in 1914 when a government fibre expert, Robert Finlow, suggested that the weed should be dragged out of the rivers and put into heaps for use as manure. However, the government was not sure at that time how far that was an economical proposition and it doubted what 'little impression would there be on the weed unless a river or khal (canal) was cleared thoroughly and the weed removed entirely'.[29] The government of India also observed that the hyacinth grew so fast that once it got started it was almost impossible to stop, and the government advised that whatever chance there might be of eradicating the weed lay in prompt action, dealing with it immediately whenever it appeared in a locality. 'In view of the danger both to material prosperity and to general health which the spread of the plant would cause', the government of India invited everyone, officials and non-officials alike, to cooperate in eradicating the pest.[30]

In spite of government decisions to destroy the hyacinth, the fibre expert retained his plan of utilization and he, along with Kenneth McLean, East Bengal's Deputy Director of Agriculture, came forward with further proposals by which the government could profit by commercial utilization of the weed. After conducting experiments in the Dhaka Agricultural Farm in 1916, they suggested that apart from high potash content, the water hyacinth was at least as rich as farmyard manure in terms of both nitrogen and phosphoric acid. In a more specific analysis, the experts-cum-bureaucrats found that the nitrogen content of the dry material was as high as 2.24 per cent and in the damp state (containing 67.8 per cent of water) it was only 0.72 per cent. Of 850 maunds (about 30 tons) of fresh green plants that were brought for experiment, about 499 maunds were heaped and allowed to rot, while the rest was spread out to be dried and then burnt. The experts observed that owing to the high water content the rotting process involved a considerable loss of nutrients. It was found that 'by drying and burning the plant the ash obtained from 300 maunds of green plant gave a larger quantity of potash than was obtained from 1000 maunds of similar plants after rotting'. The experts noted that the rotting process involved a loss of about 70 per cent of the available potash and 60 per cent of nitrogen. In other words, the key finding of the research was that burning the water hyacinth to ash was much better than rotting it in terms of nutrient value. The experts also observed that since the fresh plant contained about 95 per cent water it could not be transported economically over any distance. The rotted plant, containing about 60 per cent of water was comparable with cow-dung and it was likely that the use of the rotted material would be confined to the immediate neighbourhood of its production. But, according to these experts, the dried material was only about one-twentieth of the weight of the green plant, and was thus in a much more convenient form for transport than either the green plant or the rotted material.

Such was the spirit of commerce that the water hyacinth began to be represented as something which must be reared in earnest, certainly not destroyed. The experts reminded rural people that it was 'unwise to mix earth with the ash' and advised 'not to make ash in the rainy season, but to do in the dry weather after the middle of kartik [Autumn]' so that the plant could be 'dried for burning without fear of rain'. It was further advised that the plant should be collected from the water before it dried up in the winter; otherwise, a lot of earth would 'stick to the roots and make the ash much less valuable'.[31] It was at this time that a multinational company, Messrs Shaw and Wallace & Co., began to show a great interest in hyacinth ash.[32] The company offered the government of Bengal Rs. 4 per full unit of potash, free on rail or on board to Kolkata. The company suggested that if the ash had reached them in good condition and was not adulterated, they were ready to pay between Rs. 84 and Rs. 112 per ton. In the context of the First World War, which had restricted global access to potash, the company urged the government of Bengal to 'make it known among the agriculturists and those who can promote the scheme' and it hoped to hear from the government how the matter was received by them, and later on what progress was being made.[33] The Shaw Wallace Company, however, was not satisfied by the quality of the hyacinth supplied in the early phases of the transaction, and in 1918, the company directors informed the government of India that in future they would not buy any ash containing less than 15 per cent of potash, which was worth less than Rs. 2.4 per maund after reaching Kolkata. Consequently, the government of India advised the people: 'Do not collect any and every hyacinth that you can get hold of: but carefully select the plant. Tall, well grown plant gives rich ash and this will only be found in water so deep that its roots cannot touch the bottom such as is found in water-ways. Short leaf stalks with bulbs on them indicate hyacinth which gives poor ash and this latter plant should never be collected for making ash for sale.'[34]

Whilst the government of India, in line with the demands of Shaw and Wallace, continued to favour the cultivation of the water hyacinth, the government of Bengal recognized the danger of sustaining a policy of selective utilization of the weed. It reiterated the idea of complete destruction on the basis that while there might be a possibility of using the hyacinth as fodder, fuel, fertilizer, ash or for sale for the extraction of potash in the future, for the present all of Bengal's agriculture was under threat, and the slow pace of experiments on any of the alternatives meant that the threat would remain for many years to come. The schemes were not worth the risk of the waiting. The governor of Bengal emphasized that the danger from the weed was such that prompt eradication seemed to be 'the first consideration and that the question of its utilization ... must give place to that of its complete extinction'. He suggested that it was the duty of the local bodies (district boards, local boards, union committees and the municipalities) to eradicate the weed by all means in their power 'whether or not arrangements could be made to use the plant profitably'.[35]

At about this time the seven-member Water Hyacinth Committee was appointed by the government of Bengal with Sir Jagadish Chandra Bose, a renowned Bengali botanist, as president. The committee held seven meetings between August 1921 and August 1922 before publishing its report. In this it was observed that the districts of East Bengal, except Chittagong and the Chittagong Hill Tracts, were 'all badly infested'. Considering the extraordinary rapidity at which, in some places, the plant had been spreading, the report termed it a 'public menace'.[36] The committee, however, seemed to be afflicted by the familiar dilemma of whether to destroy or utilize the weed, which dichotomy was reflected in their two main recommendations. The report suggested the undertaking of a scientific investigation 'first into the life history of the plant and its mode of propagation, and later on into the practical methods for its check, and the economic utilization of the hyacinth in various ways so that the cost of operations may, to a certain extent, be recovered'. For this purpose, it was recommended that a plant physiologist, a subordinate officer of the agricultural department and an agricultural chemist be appointed for three years. On the whole, the committee seemed to approach the water hyacinth mainly as an object of scientific experiment, with little sense of urgency and in the hope, eventually, of recovering the cost of experimentation from its commercial utilization.

With this background, it is little wonder that scientific research tended to concentrate more on inventing methods of utilization of the weed than on finding ways to challenge its growth. By the time the debates about scientific means of dealing with the weed – for instance, whether the growth of the water hyacinth took place through the seeds or stem – faded in the 1920s, H.K. Sen, Ghose Professor of Applied Chemistry in the University of Calcutta, had started experimenting on methods by which it could be utilized. Around 1930, Sen claimed that, as with maize-stalks (Mazolith) and woodchips which were widely used in America, forming solid blocks of materials out of the hyacinth might similarly prove productive. Sen envisioned that before the air-dried weed was brought to the plant for converting into manufactured products, over 150 agriculturists and peasants could find work for every 100 maund-a-day plant. At a later stage of production at each such factory, 50 young men could find employment. Considering that about 4269 square miles were covered with the water hyacinth, quite a large industry might be established. According to Sen, it was possible to remove the plant to different areas from time to time and the rate of Rs. 1.8 per maund should be sufficiently attractive for the cultivator. He also suggested that alcohol could be made out of this weed.[37]

Meanwhile, B.K. Banerjee, a contemporary commentator, identified and compared available methods of eradication of the hyacinth, namely 'biological', 'mechanical' and 'chemical or thermal'. Banerjee did not favour the biological method on the grounds that no biologist anywhere in the world had been able to discover either a fungus or suitable bacterium or an animal or a plant that could destroy or even contain the water hyacinth. With respect to mechanical means, Banerjee calculated that a labourer could destroy 800–1000 square feet of weed per day, at a daily wage rate of between six–eight annas (1 anna equals one-sixteenth of a rupee). Banerjee also noted that the mechanical solution might lead to coercing the labouring class into clearance activities, depriving them of their daily earning from their own agricultural work. In comparison to the first two options, Banerjee found the chemical or thermal method better on economic grounds as well as in terms of effectiveness. He referred to one Subimal Bose who had invented a 'spraying solution' that killed not only the floating vegetative parts of the weed but also the stem, which remained beneath the surface of water. According to Banerjee, Bose's spraying solution cost about two annas per gallon – possibly less, if large-scale production were arranged – and since one gallon was 'sufficient to destroy completely the weeds covering an area of 300 to 350 square feet', the cost of clearing 900 square feet came to about six annas. Keeping all these factors in mind, Banerjee found the spraying solution a 'most satisfactory way of grappling with the problem of eradication'.[38]

By the mid-1930s, in spite of several attempts informed either by an honest intention to deal with the weed or by a desire to make profit out of it, there had been neither a breakthrough in scientific means of destruction nor in industrial or any other form of utilization of the water hyacinth. In the face of claims that several chemical sprays had the power to destroy the weed, some of these materials were examined by the Water Hyacinth Committee, notably by Griffiths, a South African scientist, and another Bengali chemist; but none of the claims of effectiveness of the sprays could be proved. At the same time, an institutional incapacity also surfaced. At the conference of the Union Boards of Dhaka in July 1933, the governor of Bengal, John Anderson, conceded that it was 'abundantly clear' that eradication could only be achieved by 'simultaneous attack over the whole field of operations'. But he noted that the Department of Agriculture and Industries, under whose purview the issue of the water hyacinth lay, had not the machinery, even if a method could be agreed upon, to carry out such a large scale campaign against the hyacinth throughout the province.[39]

After 1936, with the introduction of the Water Hyacinth Act, an opportunity arose for legislative action towards eradicating the weed. However, two long decades had elapsed between the opening of the destruction/utilization debate and the initiation of formal legislation to combat the hyacinth in Bengal. Legislation had been passed in Cochin China (1908), in Burma (Water Hyacinth Act of 1917), in Madras (Agricultural Pests and Diseases Act of 1919) and in Assam (Water-Hyacinth Act of 1926). The question, therefore, arises as to why it took such a long time to reach this point in Bengal, and how the legislation, when introduced, impacted on the agrarian Bengal Delta. The following sections focus on these issues.

Towards legislation

In 1919 the government of Bengal made enquiries about the legislation on the water hyacinth that had been introduced in Burma, with a view to adopting a similar legislative measure in Bengal. After analysing the reply from Burma, McAlpin, Secretary to the Department of Agriculture in Bengal, found that the Burmese government had in fact abandoned any plan for total eradication and had confined their action to keeping open the main waterways. In private circles McAlpin termed the letter from Burma a 'blow', and felt that the Burma Water Hyacinth Act was a failure. He suggested they 'had, therefore, better say nothing about it'. McAlpin felt that if an act for total eradication had been a failure in a province where the government had greater executive powers than in Bengal, such an act in his own area would 'most probably be quite useless'. He, therefore, suggested dropping the question of legislation. The file was then sent to the governor for cancellation of the programme. The personal secretary to the governor, referring to the probable consequences of this development, noted: 'I am afraid this is going to be worse even than the rabbits in Australia!'[40]

While the first attempt to introduce legislation on the water hyacinth was more or less abandoned before it got started, in January 1921 a by-law was framed and approved at a conference held in Dhaka. The Dhaka conference resolved that legislation was the only way to contain the water hyacinth. However, similar by-laws were not introduced in other districts, except sparingly in a few sub-districts; nor was the government ready to legislate the issue of eradication of the water hyacinth on a comprehensive scale all over the province. It was reported that the government was awaiting the result of the working of the Dhaka by-law before committing itself to any form of legislation.[41] Meanwhile, the Dhaka by-law itself was far from being operationally perfect. Apart from being localized in nature, the by-law was weak, as it did not provide for notices for clearing to be issued more than once a year. The Water Hyacinth Committee itself reported that the Dhaka by-law failed in that it only stipulated clearance of the weed once a year although experiments had shown that at least two clearings were necessary within a short interval, as there were generally a number of plants missed in the first clearing. It appeared that even if there were clearing operations more than once a year or even once a month, the situation probably would not have improved, as was reflected in the statement of some of the delegates of the Dhaka conference who were against the very idea of local legislation. They argued that the hyacinth affected each district differently, and that it was difficult to impose penalties on individuals who claimed that their land was invaded by the hyacinth from upstream or from another district. It was agreed by the delegates that district boards were powerless unless an act was introduced and applied all over India.[42]

The Water Hyacinth Committee prescribed that 'some form of legislation should be adopted which will ensure that concerted action is taken when applying methods designed to destroy the weed'. However, it became apparent from the minutes of the meetings of the committee that it was not easy to translate these recommendations into reality. The wording of the recommendations implied that legislation would follow the invention of scientific methods of eradication. Politics raised further questions, in that, as argued by Sir Jagadish Bose himself, any kind of legislation could be misunderstood and antagonize the people, while owing to the poor state of funding, the government would not be able to aid them. Another member of the committee, S.N. Sufi, remarked that they could not penalize the public unless the committee could tell the sufferers the best way of eliminating the weed. He warned that their best intentions might be thwarted by the fear that they were simply going to introduce a new mode of taxation without doing anything particularly useful.[43] Though the committee members felt legislation would be politically awkward, they nevertheless recommended 'some form of legislation', not, it should be noted, a comprehensive legislation. This of course was the wrong line of action since, given the pattern of spread of the hyacinth, only a comprehensive inter-district and inter-provincial effort could be successful in eradicating it. At the same time, the committee, though aware of its practicality, did not recommend frequent and regular destruction of the weed. In the case of French Cochin China landlords and tenants were obliged to clear the weed during the first three days of every month, although the authorities in French Cochin China failed to apply the regulations rigorously. The idea of monthly clearings was never taken up in the by-laws and regulations in Bengal and the question of legislation was further held up in the wake of the economic depression of the early 1930s.[44]

The debate about the ways and means of dealing with the hyacinth continued, particularly in relation to the recommendation of the Royal Commission on Agriculture in India. In their report, the commission recommended that the problem of the water hyacinth in Bengal should be dealt with by legislation similar to that which had been enacted in Assam, Burma and Madras. It doubted, however, whether legislation prescribing the destruction of the hyacinth, or measures to prevent its spread such as the construction of storage pounds or floating fences, would prove more than palliative. The commission, therefore, recommended that the formulation of a programme for research on this weed should be the top priority of the proposed Council of Agricultural Research. The government of India favoured the second of these recommendations.[45]

While the question of legislation was shelved as a matter of secondary importance, the prioritized scheme of research on the weed surprisingly failed to include Bengal whereas Bihar and Orissa, where the problem was much less acute than in Bengal, was given more attention. After examining the papers sent from the three provinces, the council determined that the situation in Bihar and Orissa demanded action, but with respect to Bengal it came to the conclusion that 'no action was required on the part of the Council'.[46] The Bengal Waterways Act, which was passed a few years later, made only a passing reference to the problem of the hyacinth. The act, passed in 1934, suggested the formation of a Waterways Board, which could clear or destroy the weed in any district where there were 'navigable channels under the control and administration of the Board'. This meant that only the water hyacinth in large 'navigable channels' came under the board's jurisdiction.[47]

The first all-Bengal legislation was passed in 1936. The act provided for some tough measures in the case of failure to eradicate the weed. In some ways, the legislation appeared to be too tough and difficult to sustain for ordinary people. By this act, the collector of a district was empowered, if he failed to recover the cost of eradication, to enter and take possession of any land or water bodies at his discretion. He could do so when costs were due, and he had the power to retain possession of the land and 'turn the same to profitable account until the said costs together with interest thereon' could be realized from the profits or paid by the occupier. The ceiling of interest was fixed at 6.25 per cent.[48] The act also stipulated that the amount spent by the collector in the course of eradication of the weed would be 'recovered from the persons benefited with interest'. Beside the question of interest, this legislation made one thing clear: the government took no responsibility for the practicalities of eradication, which now rested firmly on the occupiers of affected land,[49] although the Water Hyacinth Committee of 1921 had warned against such a measure, that is, it had warned against legislating without showing how to eradicate the weed. Neither did the act state specifically when and how many times a year clearing operations had to be undertaken. Then there was the problem of violating people's private space. The act empowered the collector of a district to occupy land for the purpose of destruction of the water hyacinth for as long as six months. For compensation, it was suggested that if any material damage or injury was caused thereby to the occupier of such land, the collector shall 'pay to him such compensation as shall be agreed upon in writing between the Collector and such occupier; provided that in assessing such compensation the manurial value of water hyacinth destroyed thereon shall be taken into account'. Once again, the idea of eradication, as envisioned in the legislation, was compromised in that the notion of the possibility of commercial uses of the weed was left intact. In clause 18, it was stated that notwithstanding anything contained elsewhere in the act, any person or class of persons, authorized by the local government, might 'sell, remove or keep water hyacinth for a prescribed purpose'.[50] The legislation of 1936 thus actually secured the idea of utilization of the water hyacinth.

Legislation and beyond

It seems unlikely that even if the act of 1936 had been better crafted, drives for eradication of the water hyacinth would have been successful, not least because the government failed to prioritize the issue of containing the weed within its schema of governance. For instance, instead of launching an all-out drive against the hyacinth following the legislation, the government decided to wait until the results of research that was being carried out in Orissa under the auspices of the newly formed Council of Agricultural Research. The government hoped that the research would produce sufficient new materials to justify a re-examination of the problem of the water hyacinth.[51] In June 1938, the agriculture minister, Tamizuddin Khan, informed the Legislative Council that an accurate estimate of the area covered by the hyacinth throughout the province of Bengal would require considerable time and expenditure and that a comprehensive drive for eradication was not considered necessary.[52]

In the last week of April 1939, a 'Water Hyacinth Week' was launched by the government in an effort to start a 'concerted and simultaneous drive' to eradicate the weed all over the province. This appeared to be the best effort on the part of the Muslim League-Krishak Praja coalition government to meet its election pledges, which had included an assurance of the eradication of the water hyacinth. Water Hyacinth Week encouraged a festive mood: civil servants were mobilized, ministers moved into every corner of the countryside, and people in general joined hands – all in the name of eradicating the water hyacinth. Students were advised to form boat racing clubs in the hope that once established, members of the club would have the 'double enjoyment' of not only participating in boat races but also of clearing the weed wherever they appeared. In some areas boys were encouraged to kill as many snakes as possible since these often hid in the thick mat of the water hyacinth. In Dhaka, a 17-year-old boy was promised a gold medal for bagging most of the 64 snakes killed during Water Hyacinth Week. The girls did not lag behind in the race and the chief minister of Bengal, H.S. Suhrawardy, himself acknowledged that the work done by some of the schoolgirls in Bogra district was 'even better than the results achieved by the boys'.[53] Observing the enthusiasm of the Scouts, schoolboys, pundits, maulovis, peasants, landlords and lawyers in Kishorganj, Suhrawardy hoped that in 'fighting common enemies like water hyacinth, there should be no difference between the different communities' and that the 'healthy teamwork was bound to destroy all Hindu Muslim quarrels'.[54]

At the end of the week, one English civil servant was reported as having 'sun-stroke' and another of being 'stuck in the mud',[55] but no long-lasting solutions to the problem of the water hyacinth were in sight. No doubt considerable areas were cleared of the weed, but as time passed, the orchestrated enthusiasm faded away: the ministers returned to Kolkata, the officials went back to their mofossal headquarters and the schoolchildren to their classrooms. Those peasants and villagers who were actively engaged in agrarian activities continued to face the same water hyacinth problems. To celebrate a Water Hyacinth Week might have been an astute political move by a ruling party, but its failure was inevitable, because the problem was also biological and environmental in nature, which demanded an examination of the changes in the ecological system that encouraged the growth of the plant. These issues were indeed raised. Two weeks before the Water Hyacinth Week was launched, Sudhir Chandar Sur opposed the idea, which he thought was intended to remove the water hyacinth without treating the causes of its growth. Sur attributed the growth of the weed to the obstacles to the current of rivers and other watercourses posed by cross-roadways, railway embankments and the pillars of the railway bridges. Sur argued that such obstacles prevented different waterways from performing their natural functions of clearing away large amounts of organic matter to the sea via bigger rivers. This resulted in the deposition of organic matter in the beds of the watercourses and the water hyacinth found a congenial environment there. However, Sur felt that compared to the long-term impact of the blockage of water currents, the effect of the water hyacinth was minimal. Sur even suggested that the water hyacinth was beneficial for the time being since it consumed organic matter, preventing many parts of the delta being transformed into marshes full of animal organic matter. In this context, Sur thought that the water hyacinth would be welcome until it threatened entirely to choke up the already dying watercourses of Bengal. He suggested that the water hyacinth itself should not be tackled unless the artificial agencies, which had reduced the water currents in big rivers, had been tackled first, since stronger flow in the water bodies would automatically lead to the clearance of the weed.[56]

There is no denying that the Water Hyacinth Act of 1936 reflected a growing consensus on the importance of getting rid of the weed and concern for the agro-ecological future of the Bengal Delta. What seems important in this context is to examine how this consensus was informed and articulated by different competing forces in society and the state. In many cases, local efforts were frustrated by the lack of cooperation and coordination between the government and common people as well as between different government departments. For instance, it was alleged that in the Arial beel areas in Munshiganj of Dhaka, about

50,000 flood-stricken cultivators had invested substantial borrowed capital with the encouragement of a certain local government officer. But the cultivators were on the brink of disaster as no initiatives to implement a promised water hyacinth control scheme had taken place. When this was referred to in the legislative assembly, the minister for agriculture noted that it was not a government scheme but was 'suggested, worked and paid for by the local people with the assistance of a Special Officer'. The scheme was specifically aimed at constructing a barricade across the waterways surrounding the low lands of the beel in order to check the spread of the hyacinth, but the Speaker of the Assembly denied any government responsibility regarding this and remarked that the construction of a barricade rested entirely on the local people. He did not elaborate why, in such circumstances, the peasants would resort to agitation.[57] In another instance, while it was claimed by the provincial government of Bengal that the act of 1936 was introduced to empower the district authorities, land belonging to the railway authorities was not covered as they fell under the control of the government of India.[58] Since railway and roadside ditches and waterways blocked by railway embankments were commonly places of regeneration and growth of the hyacinth, the exclusion of these lands from the jurisdiction amounted to a technical farce as far as the programme of eradication of the water hyacinth was concerned.

An amended Bengal Water Hyacinth (Amendment) Act, 1940, empowered an authorized officer to prepare a scheme of any work relating to the water hyacinth and to realize the cost for such scheme proportionately from the benefited persons. There was, however, no provision empowering the authorized officer to realize the cost of the removal and destruction of the water hyacinth, which could be intercepted in any common flowing channel as a result of the execution of such a scheme. Therefore, instructions were given to the authorized officer to be 'so good as to take every care in the execution of schemes under section 3 of the Amendment Act so that no water hyacinth is intercepted in any flowing channel'.[59] There was also the problem of intra-governmental coordination in the whole project of combating the water hyacinth. A special officer, who was appointed to deal with the water hyacinth, noted that work against the weed, including local clearance and the setting up of barriers in key positions, could not be implemented properly because of differential administrative arrangements. The officer observed that government works relating to water supply or the setting up of dispensaries were done more or less by respective departments independently, but this was not the case with the water hyacinth. In terms of dealing with the water hyacinth problem there was no contractor to carry this out in anticipation of payment and there was no organized agency to help.[60]

Given the varied and often self-seeking response to the problem of the water hyacinth by different agencies within society and the government, the legislation and apparent consensus to destroy the weed was found to be ineffective in many ways. The lack of genuine efforts to tackle the problem was amply matched by the lack of focus within the policies and programmes of local political forces. Referring to the fact that there was an unthinkable hahakar (a widespread hopelessness) and tremendous poverty in Bengal due to the growth of the 'bloody plant', a Bengali newspaper commented:

> The rural inhabitants of Bengal have gradually become sick and idle. There is no enthusiasm, nor encouragement or initiative among them. They don't try to destroy this enemy [hyacinth]. They are sitting idle thinking that this is a curse from God. If some day God himself withdraws the weed, only then their lands would be free and the mouths of the rivers be opened. This class of fatalist cowards even dreams of *swaraj*![61]

The water hyacinth thus survived the wrath of the Bengal Chamber of Commerce in the 1910s, the scientists' chemical spray in the 1920s, and electoral commitment, legislation and above all a historic 'Water Hyacinth Week' in the 1930s – all of them aimed at its destruction. For a tiny, relatively weak aquatic weed, 90 per cent of which comprised harmless water with a tinge of 'feminine beauty', this has been no mean achievement. But its survival, along with the railways, has had serious implications for the economic well-being of the agrarian population, which had no lesser connection to the great Bengal famine – a topic that we will deal with in the next chapter.

8
Between Food Availability Decline and Entitlement Exchange: an Ecological Prehistory of the Great Bengal Famine of 1943

One autumn morning in 1906 in an East Bengal village, Sister Nivedita[1] came across a few women standing up to their throats under water, gathering unripe grain stalk by stalk. As she offered these women her boat and assistance, they said they could not accept because they were naked. Nivedita was in East Bengal on a famine relief mission; hence she came across many such incidents and was no longer shocked by different degrees of human suffering. But she was surprised by the 'freshness' exemplified in these women's experience of destitution and starvation. She noted: 'Since my visit to Eastern Bengal I have had the opportunity of comparing the people ... with those of another district nearer the capital [Calcutta], where famine and destitution have of late years become chronic. And I have learnt thus to measure the freshness of impression of hunger by the shrinking from loss of personal dignity in the stating of need.'[2] Nivedita further observed that everything that one saw in East Bengal that day was 'so much saved from happier times' and warned that if the present strain continued long enough, it would surely give way to a 'sordid pauperism'.[3]

Nivedita's account supports our contention of relative well-being in nineteenth-century East Bengal and points to the fact that the great Bengal famine of 1943–44 has an under-studied prehistory.[4] Of about 1.714 million people who died in eastern Bengal because of famine or famine-related diseases,[5] most were landless agricultural labourers and the members of other vulnerable groups, including fishermen, artisans, and the permanently workless destitute and beggars. Any discussion of the Bengal famine therefore needs to take note of the historical conditions out of which these people emerged. Amartya Sen, a leading authority on the Bengal famine, has been criticized for not focusing on the long-term agrarian-historical changes that climaxed in the famine. It is suggested that although Sen perceptively employed the theory of entitlement exchange[6] to explain the famine, he did not pay due attention to the historical construction of the group of people who had partial or no entitlement to food or to a remunerative labour market. While this is too much to expect from an economist whose focus was on the immediate causes of the famine, such criticism has nevertheless encouraged studies that examine the famine from longer-term perspectives. Economists with historical interests have particularly focused on trends in food production, rising rates of taxation and macro-economic downturns resulting from the great depression of the 1930s.[7] Social histories have tended to relate the economic interpretation to the emergent cleavages within agrarian society that widened the gap between the rich and the poor, also in the context of the great depression. There is scarcely any truly long-term environmental-historical perspective. Even those who defended the food availability decline (FAD)[8] approach, as opposed to Sen's entitlement approach, focusing on the fall in food production as a result of cyclones or flood can be similarly criticized for not considering the accumulated effect of longer-term environmental problems, that is to say, those whose causes lie more than a couple of years prior to the famine.[9]

The previous three chapters have drawn a broad picture of ecological deterioration (Chapters 6 and 7) and the complex political-ecological relations that eroded cultivators' entitlement to land, and which contributed in the early twentieth century to the emergence of a pauperized rural population (Chapter 5). This chapter examines the wider ramifications of those developments in the context of the great Bengal famine. It will be seen that the FAD and entitlement approaches are not as mutually antagonistic as might appear from the fierce debates between their respective proponents. In fact we may get a clearer picture of the causes of the famine only when a long-term environmental-historical view enables us to appreciate these two approaches simultaneously. We will attempt to achieve this here, first, by examining the different and long-term ecological contexts of FAD and, second, by extending Sen's entitlement approach to longer-term changes in the social relations of land ownership and production. The chapter ends with a note on the impact of ecological changes on health and nutrition and the colonial state's response to the problem, in an attempt to understand the extensive disease-related famine mortality that accompanied deaths from starvation.

Ecology and FAD (food availability decline)

The first decade of the twentieth century saw Burmese rice being imported into India in unprecedented quantities. Whereas between 1881 and 1890, 64,000 tons of Burmese paddy and rice were imported by India each year, between 1891 and 1910 imports reached 571,000 tons, rising to 797,000 tons per annum from 1911 to 1920.[10] Dependency on imported rice from Burma increased over the decades to such an extent that the stoppage of its import in the wake of the Japanese invasion of Burma was considered to be one of the major causes of the Bengal famine in 1943. Scholars maintaining a short-term intensive focus on the Bengal famine have not asked why, in the first instance, Bengal had been dependent on imported rice for several decades before the famine. A satisfactory answer to this question could clarify some key issues regarding FAD.

There appear to be two major factors underlying food availability decline: under-exploitation of cultivable land and decreasing trends in crop output. During the nineteenth century, as we have seen, the Bengal Delta saw an enormous pace of reclamation and cultivation of wasteland. In Barisal, for instance, between 1860 and 1905, 'unoccupied waste' shrank from 526 square miles to 184 square miles. One hundred and eighty square miles of new land was brought into existence by the fluvial action of rivers, and cultivation increased by about 23.1 per cent in the occupied area.[11] However, from the beginning of the twentieth century, a paradox unfolded. In 1931, the population density in East Bengal reached between 900 and 1200 persons per square mile,[12] which called for an even more aggressive process of land utilization. In reality, as Table 8.1 shows, the amount of cultivable land remained more or less static in spite of remarkable population growth and corresponding demand for food.

In addition to the impact of static or under-exploitation of cultivable land, an important underlying factor of FAD was the gradual fall in production output in East Bengal. Between 1906 and 1937 the average yield of aman, the main rice variety in the region, showed steady decline (Table 8.2).

A number of studies on Bengal agriculture support the above data. M.M. Islam shows that there was hardly any improvement in yield in Bengal at the aggregate level from the early years of the twentieth century and that there was only a marginal expansion of the acreage under cultivation, which represented a 'marked disparity' between low expansion of agriculture and population growth.[13] Although Islam thinks that

Table 8.1   Cultivable wasteland (other than fallow) in 11 districts of
the eastern Bengal Delta

Year	Acres
1900–01	1 556 576
1911–12	1 113 248
1915–16	995 730 (excluding Pabna)
1919–20	1 048 222 (excluding Faridpur)
1925–26	1 101 831 (excluding Pabna and Faridpur)
1930–31	1 509 988 (excluding Faridpur)
1935–36	1 682 175 (excluding Faridpur)
1942–43	1 500 769

Note: The eleven districts are Barisal, Chittagong, Dhaka, Faridpur, Jessore,
Khulna, Mymensingh, Noakhali, Pabna, Rajshahi and Tippera.
Source: Agricultural Statistics of Bengal, 1901–02 to 1942–43 (Calcutta), p. 28.

Table 8.2   Average yield (lb per acre) of aman (winter rice)

District	1906–07	1921–22	1936–37
Barisal	1344	1047	981
Chittagong	1008	1156	958
Dhaka	1120	935	959
Faridpur	1232	1017	974
Jessore	1545	1022	889
Khulna	1361	1020	1018
Mymensingh	1120	934	926
Noakhali	1008	933	787
Pabna	1008	957	956
Rajshahi	1008	988	831
Tippera	1008	914	962

Source: Agricultural Statistics of Bengal, 1901–02 to 1942–43 (Calcutta).

the 'virtual stagnation' in the all-crop acreage was understandable in
the context of the limited scope of extension to new areas, he is curious as to why there was no significant improvement in yield per acre
despite a relative abundance of labour. Suggesting that the cultivators
might have taken rational production decisions such as were desirable
within the given technological and institutional constraints, Islam
nevertheless believes the real explanation of the low productivity and
near-zero trend between 1920 and 1946 seems to be the 'low level of
capital formation'.[14] These are important findings from the perspective
of technological modernization and capitalist agricultural development;
but these conclusions are drawn without any particular attention to

the ecological conditions prevailing in Bengal. Such a purely economic explanation fails to explain why, for instance, in the late 1920s a family was taking meals which consisted entirely of wild herbs and snails gathered from a nearby marsh, although it possessed about 20 acres of good land.[15]

Many of the ecological problems that can be connected to under-exploitation of land and under-production of crops have been described in the previous two chapters. In what follows we will look at agrarian ecological conditions during the crop year immediately following the famine. For this purpose, I will mainly depend on an authoritative official report known as a 'plot-to-plot' survey.[16] These statistics and empirical findings are important because they reflect the continuation of ecological problems prevalent from the start of the twentieth century through to the famine years. The data were collected during the two years immediately following the famine, when the 'grow more food' campaign was under way, hence offering a picture of environmental conditions which were better than those of the famine period itself.

Based on data presented in the report, the districts of eastern Bengal were arranged into two broad groups. The first group consisted of those districts where cultivable fallow increased or was in a static condition. The second group represented the districts where cultivable wasteland decreased. The comparisons are measured by the data provided in the plot-to-plot survey and the settlement reports published about two decades earlier and included in the report.

The first group (in which cultivable fallow increased or remained static) comprised the districts of Jessore, Comilla, Pabna, Faridpur, Barisal and Noakhali. Among the common problems in these districts were the water hyacinth, inadequate drainage of waterlogged areas and the drying up or silting up of rivers and other water bodies. In the Jessore district all subdivisions were 'badly affected' by the water hyacinth, whereas in Pabna standing crops were destroyed by the weed. In Comilla, the water hyacinth rendered large areas, particularly the beel areas, 'unfit for cultivation'. Local people believed that if barricades to stop the water hyacinth had been erected, 'thousands of acres of fertile paddy lands' could have been saved from damage and destruction. Faridpur and Barisal had similar problems as far as the water hyacinth was concerned.

With respect to waterlogging, in Jessore general productivity and yield per acre declined significantly as a result of a lack of proper irrigation and drainage as well as deteriorating health conditions of the cultivators. In Pabna, standing crops were destroyed by waterlogging and other water-related problems connected to the railways, particularly the Sara-Sirajganj railway. In Comilla, the river current was slow and there was loss of fertility in the land as less silt was deposited on the paddy fields. In Faridpur, reports of waterlogging observed that it had made 'soil less fertile, transport of produce more difficult and the climate more unhealthy'. It was believed that this was partly due to the gradual eastward movement of the Ganga and the construction of the railway lines and bridges in the north interfering with the region's drainage system. In Barisal, there were no railways, but problems were created by protective or road embankments. Besides, there were more than 8000 acres under water hyacinth. In Noakhali, the water hyacinth was not the most serious problem, but the death of rivers was. In some places premature settlement and attempts to protect crops from saline water penetrating embankments had created permanent depressions and waterlogging in the interior of the district.

The second group of districts (which showed signs of decrease in cultivable wasteland) included Khulna, Dhaka, Rajshahi and Mymensingh. In Khulna, this was possibly because of the lesser extent of the impact of the railways and the water hyacinth. According to the report, in this district the growth of the water hyacinth was checked by the saline coastal water, implying that the problem of water hyacinth was replaced by that of saline water. However, in spite of the decrease of cultivable wasteland, that is to say, the equivalent expansion of land under cultivation, agricultural productivity did not increase, largely as the result of other ecological problems, particularly relating to coastal embankments and the prevalence of diseases such as cholera and smallpox. In Rajshahi, a considerable amount of land was reclaimed from the receding Chalan beel. But this did not translate into increased crop output as the Chalan beel was widely affected by the growth and spread of the water hyacinth across at least 10,000 acres. There were complaints that considerable damage to crops resulted from the inadequate number of culverts through the railway embankments, which could lead to either lack or excess of water in the fields. (See Figures 8.1 and 8.2.)

In Dhaka, a considerable amount of land was created at the expense of dying rivers and some of these areas came under cultivation. But increased productivity was not possible in the newly-formed lands, because most areas of the district were affected by the water hyacinth. The decrease of cultivable wasteland in Mymensingh was primarily the result of a considerable increase of migrants into the region. Since Mymensingh was en route to Assam's Brahmaputra Valley, the lands along the way were cultivated by the settlers in the Mymensingh side of the valley. Yet, again, the decrease in cultivable fallow did not lead to

Figure 8.1 Embankment, 1930s (i)

Note: The photo shows a reclaimed area in the Sundarbans. To the left is Harda Khal, a tidal channel and on the other side of Harda Khal the land is still under the natural mangrove forests. The embankment in the middle of the picture protects the cleared land from tides. On the cleared area, the incidence of *Anopheles ludlowii* mosquito is very heavy, whereas in the afforested areas, it is entirely absent.

Source: M.O.T. Iyengar, 'The Distribution of *Anopheles ludlowii* in Bengal and its Importance in Malaria Epidemology', *IJMR*, XIX, 2 (1931), plate XXX.

a corresponding output in agriculture. While the railways destabilized the Brahmaputra water system, most subdivisions were seriously affected by the water hyacinth and waterlogging. In the bhati areas, owing to the 'bottle-necking' of the Meghna by the Bhairab Bridge the huge quantity of water that came down from the Sylhet and Garo hills did not drain quickly enough and consequently aman cultivation became 'extremely precarious'. It was these ecological problems that resulted in Mymensingh, which saw the largest rate of the shrinkage of cultivable wasteland in eastern Bengal, failing to attain any corresponding agricultural recovery.

In total, during the crop survey of 1944–45, the water hyacinth is recorded as affecting 269,000 acres of cultivable land.[17] This was at the same time as the 'grow more food' programme was under way, implying much better agro-environmental conditions than those of the famine years. One or two examples of the conditions relating to the 1942–43 crop year may be pertinent here. A government note of 1943 pointed out that as the result of the ravages of the water hyacinth, large tracts

Figure 8.2   Embankment, 1930s (ii)
Note: The photo shows a village in the cleared area of the Sundarbans showing the high embankment on the right constructed to protect the land from the tides. The river is to the right side of the embankment. Anopheles mosquito breeds heavily on such cleared and protected areas.
Source: M.O.T. Iyengar, 'The Distribution of *Anopheles ludlowii* in Bengal and its Importance in Malaria Epidemology', *IJMR*, XIX, 2 (1931), plate XXX.

of lands had been converted into fallow lands and crops of the value of 'lakhs of rupees' were destroyed by this pest. It was further noted that the value of crops saved by two eradication schemes in Dhaka could well be more than 10 lakh rupees annually.[18] A report published in September 1942, regarding the crop prospects for Natore of Rajshahi district, revealed that 'total destruction' of aman was feared if the water hyacinth was not removed quickly from several thousand bighas of paddy land in beel areas and many village fields.[19]

While the major results of flooding and excessive waterlogging were crop failure and under-exploitation of cultivable lands, these problems also contributed to various crop diseases. Urfa, a rice disease, came to light at the beginning of the twentieth century. In 1930 a disease called rice hispa (from *Hispa aenescenes*, a small black beetle) ravaged rice fields in Barisal to such an extent that it was necessary to distribute agricultural loans over 200 square miles.[20] The same year in the same district, there was a report of a fungus, *Helminthosporium oryzae*, commonly known as 'brown spot' disease, which damaged a considerable amount of paddy.[21] In 1938–39, the disease of *pan-sukh* was added to the list. All these diseases were related to some degree to the poor drainage condition in the rice fields.[22] Any one of these could have struck Bengal on a wider scale from the 1930s onward, but it was *Helminthosporium* which attacked the harvest in 1942. According to Padmanabhan, who made the first scientific examination of the relationship between *Helminthosporium* disease and FAD, one of the essential conditions for attracting the disease was waterlogged ground.[23] Padmanabhan has shown that climatic conditions in Bengal were perfectly suited for the disease in 1942–43.[24] A combination of meteorological conditions and waterlogging encouraged the infection, which resulted in yield losses estimated at between 50 and 90 per cent in the 1942–43 crop year.[25] It was asserted by Padmanabhan that nothing as devastating as the infection of this particular year had previously been recorded in plant pathological literature. The only other instance that bears comparison in terms of both the loss of a food crop and the human tragedy that followed was the Irish potato famine of the 1840s.[26]

Thus, the worsening ecological conditions in Bengal explain the static or disproportionately low exploitation of available cultivable wasteland and lower aggregate crop output more or less from the start of the twentieth century. It was not surprising that the 'normal' deficit of rice in Bengal stood at 64,000 tons in the few years prior to the famine.[27] This is where the real FAD question arises, and these facts also lend support to the Foodgrains Procurement Committee's findings that net decline in the Bengal 'carry over' during the seven years up to, but excluding, the famine year, might well have been over 2.5 million tons.[28] It is no wonder that millions of people – even in normal times – subsisted only just above starvation level.[29] In this context it is understandable that shortfall in aman production in 1943 was neither as 'moderate' nor as 'indifferent' as Amartya Sen suggests. Mufakharul Islam notes that the decline was 30 per cent compared to 1942 and 20 per cent compared to the previous quinquennium.[30] This aspect of ecology-induced FAD, which reflected the general long-term declining trend in rice production in Bengal, must be kept in mind in any assessment of the causes of the famine.

De-entitlement at work

Even if we agree, in the context of our previous discussions, that ecological problems contributed to the increase of fallow land and the decline of crop outputs, leading in turn to FAD, this does not necessarily displace the entitlement exchange approach. Low productivity or lack of productivity does not in itself explain famine; we must at the same time link up differential individual capacity to access food within the constraints of FAD. Sen argues that food was available, but that a certain group of people did not have access or entitlement to it. Sen is correct about entitlement failure in terms of loss of purchasing power in the market place, but he is less interested in the specific problem of the historical construction of de-entitlement in relation to land.[31] At the same time, although some critics argue that Sen's assumption is not new, in no such critical discussions is the emergence of the vulnerable groups adequately traced.[32]

Chapter 4 considered the intricate agrarian relations out of which a group of land-poor and landless vulnerable people emerged. It might be useful to examine the ways in which the process of a socially-constructed erosion of entitlement to land within the context of growing ecological constraints caused small-scale starvation for decades, culminating in the famine of 1943. The famine was different from earlier periods of starvation in terms of the degree and extent of casualties, not in terms of the social and ecological conditions in which it took place. The greater extent of death and devastation could be explained by the specific concurrence of events and policies between 1941 and 1943: boat denial, rice denial, cyclone, hoarding, mismanagement of the administration and so on. Yet the basic problem remained constant: the problem of entitlement to land. But how do we make the connection between the problem of entitlement to land and starvation within the strained ecological conditions of East Bengal?

For reasons described above and in Chapter 4, the number of rent-takers increased markedly in the course of the first half of the twentieth century. During the decade preceding the onslaught of the depression, between 1921 and 1931, rent-takers increased by 62 per cent while the number of sharecroppers increased by 49 per cent. By the 1940s, sharecroppers constituted 29 per cent of the total agricultural population.[33] The rising trend of land alienation or de-entitlement to land inevitably led to two other problems: first, the decreasing level of control of the peasant over the agro-ecological production process; and second, growing pressure by the neo-occupancy raiyats to extract increased amounts of rent from the under-raiyats. These issues can be examined in the context of micro-empirical data that have evaded the attention of historians and economists, who have generally dealt with macro-level data, especially relating to the market. For example, it was common practice among the new generation of non-cultivating landowners to file law suits for rent increases against their under-tenants, arguing that the prices of crops and foodstuffs and the fertility of land had increased to such an extent that a higher rent was both imperative and legitimate. In response, the under-tenants or sharecroppers argued that fertility had not increased – and consequently that they should not have to pay any increase in rent – for a number of reasons, among them: excess or lack of rain (*otibrishiti* or *onabrishti*), flooding, invasion of pests, including *ukhra, pichipoka, senipoka, ufra, kosturifena* (water hyacinth) and earthquakes. Another argument was that the neo-occupancy raiyats undermined the ecological value of cultivated areas by stopping the free flow of water by filling up or destroying different waterbodies including beel, khal, jola and nulla in pursuit of additional cultivable land.[34] The destruction of water-bodies or lack of entitlement to ecosystem services like ponds or canals caused a cultivator to lament that 'there is no *jalashai* [wetland/waterbodies] in Bangladesh ... because of the destruction of waterbodies, not only fish stock is being depleted, agriculture, human being and cattle are all in decline'. It was further claimed that in former times, peasants had the right to dig ponds on the land in their possession, but that landlords, afraid that it would result in reduced land values, were now becoming extremely hostile to such practices.[35] One victim of a law suit in Comilla, Ali Baksh, complained that crop production was half that of earlier times because of the impact of different local pests and weed – including water hyacinth – the erratic climate and declining river quality, as a result of which famine was common. Baksh argued that the prices of food and crops increased in response to low production caused by environmental problems as well as high production costs resulting from increases in the price of cattle, fertilizer and so on. Therefore the rise in crop prices was forced by circumstances, and was not a reflection of the well-being of the people.[36] But in the context of increasingly powerful political pressure groups among the neo-occupancy raiyats as well as the diminishing formative relationship between the peasantry and government, the resistance of cultivators like Ali Baksh often withered in the law courts. The ecological regime of chars, diaras and forests and other forms of khas mahals that had contributed to relatively stable social relations in the past now became the fiercest ground of contestations between the cultivators who worked the land and non-cultivating people. By the early 1920s, of Bengal's total land revenue, significantly more was coming from permanently settled areas.[37]

The gradual alienation of the state from the cultivators was becoming increasingly visible, not only because of a policy shift in agrarian relations since the turn of the century, as described in Chapter 4; this was also because the state was now more 'knowledgeable', or at least pretending to be so, about agro-ecological conditions in the deltaic hinterland of Bengal.[38] For instance, Reajuddin Sarkar of char Janajat of Comilla in his appeal to the government for retaining an existing assessment of rent noted that three-quarters of his agricultural lands were sandy and not good enough for optimum production. In response to the claim that this meant he was unable to pay his rent, the settlement officer classified Sarkar's land into 'first', 'second' and 'third' class and through this 'scientific' classification it was found that of Sarkar's total lands, most were 'first class'. On the basis of this finding, Sarkar's appeal against a higher rent was rejected.[39]

What is significant, although not unexpected, is that the government's tightening grip through new classificatory registers of the landscape, along with the problems of indebtedness and high land rental, were taking place in those ecologically endowed regions where the peasants had remained relatively prosperous throughout the nineteenth century. In these changed circumstances, the cultivators of Bengal in the khas mahals and 'particularly those of its most fertile and previously most thriving districts' became involved in debts 'far beyond their power to repay'.[40] Subject to ecological change beyond their control, cultivators were now also subject to demands for increasing amounts of rent. No wonder, the areas outside the Permanent Settlement, which had seen the rise of a vibrant peasant society in the course of the nineteenth century, were now considered nightmare zones. 'The khas mahal raiyats and raiyats in temporarily settled estates are worse off than raiyats in permanently settled estates ... raiyats cry "God save us from khas mahals". ...the zeal of the revenue officers [to collect revenue] proves a curse to the tenants.'[41]

Thus the real dynamics of agrarian relations did not reflect the impact of the Permanent Settlement; rather, it was the ecological regime of East Bengal that stood at the centre of contestations between the peasant and non-peasant elements. The historical-social construction of de-entitlement to land as well as 'surplus' extraction from the produce must be taken into consideration when talking about entitlement exchange. In other words, FAD and de-entitlement (with particular relation to land) worked in concert in East Bengal. On the one hand, land was passed to non-cultivating agencies, contributing to the emergence of land-poor or landless people. On the other hand, within the ecological constraints, relatively better lands were alienated. The poor were affected on both counts and, unable to cope with the dual pressure, joined the growing number of vulnerable groups who have been identified as the famine victims.[42]

Given the vagueness of statistics on the beneficiaries of the colossal scale of land transfer in the few decades before the famine, some historians have attributed the majority of the symptoms of rural decline to the actions of 'rich peasants' in the countryside. As we have seen, it has been argued that the process of empowerment of the rich peasant became particularly remarkable with the onslaught of the depression of the early 1930s, when the bhadralok started to leave rural areas and to engage in non-agricultural activities. The far-reaching consequence of the depression on agrarian eastern Bengal is summarized by Partha Chatterjee in the following way:

> The traditional landlord-moneylender was forced to move out and was replaced by new suppliers of credit from among the better-off section of the cultivators themselves. The usual dynamics of indebtedness leading to the transfer of small-peasant holdings was then set into full motion, helped on by the grant of full rights of transferability in raiyati lands in 1939 ... between 19 and 36 per cent of the total occupancy holdings in the districts of Noakhali, Tippera, Bogra, Dacca, Mymensingh and Pabna were sold between 1930 and 1938, much higher figures than any other district in Bengal in this period ... most of these were sales from indebted small peasants to richer cultivators ... the famine of 1943 struck what was virtually the final blow.[43]

As noted in Chapter 4, Chatterjee's observation about the 'rich peasant's' growing buying capacity and influence is correct, but in the course of the early twentieth century the category of 'rich peasant' no longer merely represented actual or original cultivators. Chatterjee's statistical observation is important from the perspective of both the causes and results of famine in terms of entitlement to land. This can be best appreciated by examining the questions relating to famine victims. Why was the mortality rate relatively greater in the towns than in the villages? What prompted the landless and poor cultivators to migrate to towns without trying to mitigate their suffering by appealing to fellow rich peasants? Such questions have to be answered from the perspective of a pseudo-urbanization that bound itself to agricultural resources after the turn of the twentieth century. During the first major twentieth-century famine in eastern Bengal, in 1906–07, Nivedita identified a 'sprinkling of intellectuals', who were scattered around the districts and the small market-towns of the countryside. They included the Brahmins, the schoolmasters, the small station masters and railway staff, clerks in village firms and shops, letter-writers, doctors and the like. Each of them was a gentleman who was 'perhaps more conscious of his pride of gentle-hood than any proud belted earl in all the West!' And for these people, according to Nivedita, the directly agricultural classes were 'the very steps on which they stand'.[44] They went on to assert more influence over the following decades, either in concert with or against the constructed animosity of the 'rich peasant'. This phoney urbanization was possible at the expense of the sharecroppers and landless in the countryside – a scenario that Gramsci identified as the 'power in the towns that auto-matically becomes power in the countryside'.[45] Calcutta represented the mega-example of this process.

Some questions relating to the problem of pseudo-urbanization can be partially examined in the contemporary famine notes of John Bell of the Indian Civil Service. After calculating that the total shortage of rice during the famine period was 'not more than 5%', Bell enquired why the famine could occur in response to such a 'trifling deficit'. Bell found answer in the battle for possession of rice between the country-men and the townsfolk. He observed that only a small proportion of the rice grown in Bengal came to market, and that what did come to the towns for sale was the surplus over the consumption of landed classes who had immediate access to and possession of food-grains. If there was a shortage, and even where there was a belief in the possibility of shortage, the cultivators and the landlords met their own requirements first, which meant a reduction of supply from the surplus for the town markets. In more serious circumstances, they would 'hold on to their stocks, and perhaps keep in their hands even larger stocks than they would normally reserve'. In some cases, they will do so with the inten-tion of realizing a better price later on, but more often with the 'human and understandable object of safeguarding themselves and their dependents'. Therefore, a modest shortfall in production was enough to constitute a very serious shortage in the amount of rice offered for sale. The immediate result was a shortage in the supply of rice for the towns, where a scramble followed to ensure supply. Prices in the town markets then rose and prices in rural areas followed suit. This confirmed the impression of scarcity, leading to more hoarding, and in turn created or intensified scarcity, so that prices rose still higher, and the process repeated itself in a vicious spiral. Since wealthier people lived in the cities and towns, they did not go short even if they had to pay more for rice. Therefore, it was the market towns that controlled the price of rice throughout the province. In such circumstances, the rentier class fortified their position not only by the acquisition of rice stocks through rental payment in kind, but also by their purchasing power. The landless labourers were, in turn, usually paid in kind for their work. When in the wake of the famine the price of rice rose, land-holders preferred to pay wages in cash, which had a lower purchasing value – or to dispense with hired labour altogether, cultivating their land themselves and with the members of their families. The labourer, finding himself without work and without food, tended to migrate to the towns, where he knew that rice was available at a price, and where he hoped to support himself either by work or on charity. This explains, according to Bell, why the mortality rate in the towns was so much higher in 1943 than in the countryside.[46]

There was little substantial difference between a sharecropper and a landless wage labourer, categories that had emerged in the decades preceding the famine from the class of under-raiyats and sharecroppers.[47] On the other hand, the majority of the 'non-cultivating owners' were well off enough to avoid selling or mortgaging their land.[48] Bell's interpretation of the role of the city and town dwellers in the making of the famine appears to have been further corroborated by the effects of the famine in the course of 1943–44. In contrast to Partha Chatterjee's suggestion, Mahalanobis worked out that of about 710,000 acres of rice land changing hands during the famine only about 290,000 had been purchased back in the villages. Roughly 420,000 acres of rice land had thus passed to outsiders, possibly, as Mahalanobis suspected, 'non-cultivating owners' residing in urban areas.[49] It was not surprising that during the famine, Cornelia Sorabji, India's pioneer female lawyer, hoped that the bhadralok would 'put their stores of Bengal rice at the service of the poor'.[50]

In the light of the particular suffering of the landless and the landpoor, the Bengal famine of 1943–44 has been rightly considered a 'class famine'. This begs the question of whether sustainable entitlement to land could have saved a substantial number of lives during the famine. If that question is even partially valid, could the extent of landlessness and land-poverty have been reduced in Bengal in the first instance and, therefore, the destructive power of the famine curtailed? In the case of eastern Bengal there was a remarkable lack of commitment to preventing agricultural lands falling into the hands of non-cultivating people. In an attempt to prevent land passing out of the hands of smallholders to non-cultivators, a Land Alienation Act was passed in the Punjab in 1901. It took until 1944 to pass a similar act in Bengal, but it was too late, for the famine had already struck.

Ecological decline, diseases and the state

If ecologically-induced lower productivity and inequitable entitlement to ecological resources were at work for several decades before their culmination in the great Bengal famine, it was nevertheless disease that completed the causal triangle. Yet the disease aspect of famine mortality has received less attention than it deserves.[51] In the Bengal Delta, malaria, in concert with cholera, smallpox, diarrhoea and dysentery, took the greatest toll during and after the famine of 1943.[52] The Bengal famine struck when the condition of public health in India was the worst in all Asia and public health in Bengal was the worst in India.[53]

Disease-related mortality during the great Bengal famine speaks of long-term changes in the public health scenario in East Bengal dating back to the turn of the twentieth century.[54] In Faridpur, malaria increased significantly from 1892 when health did not 'appear to be as satisfactory as its material conditions'.[55] In the early decades of the twentieth century malaria became almost as widespread in East Bengal as in northern and western Bengal, representing a synchronization of disease regimes across greater Bengal.[56] A plague was reported for the first time in 1898. Water-borne diseases increased remarkably from 1895–1904/05 and continued in the following decades. While between 1903 and 1909, the death rate declined in all but one province of India, it rose in eastern Bengal (along with Assam), from 31.60 to 33.89 per thousand.[57]

One way of trying to understand the worsening public health conditions in Bengal could be to examine nutritional intake. In a deltaic riverine plainland like East Bengal, the main source of nutrition was fish. As discussed in the previous two chapters, a good number of water-bodies dried up or declined following the construction of the railways and the invasion by water hyacinth, which inevitably impacted negatively on fish populations. The earthquake of 1897 caused riverbeds and beels in the Mymensingh and Rajshahi divisions to fill up. The flow and strength of the current in Ganga had decreased, and smaller branches and feeders as well as beels that were fed by it were silting up and becoming slacker. Certain methods of fishing, such as the erection of dams and some large fixed engines which obstructed the current, were also responsible to a certain extent for raising the beds of the water-bodies. Small-span culverts and bridges, whether on railway lines or on highways, also helped to weaken the current and to deposit silt. The project for a railway bridge at Sara Ghat on the Ganga raised fear among the fishermen that it would slow the current and thereby contribute to siltation. The contamination of breeding grounds in shallows and back-waters by the retting of jute also caused fish to die. It was reported that steamers running through narrow and shallow water routes destroyed the ova by their wash, scaring away shoals of fish in their migration to their breeding-grounds and attracting and blinding the fish with their searchlights.[58] At the same time, the construction of the railway resulted in the depletion of indigenous fish stocks and deterioration in the conditions of the fishing community, which was much worse off than had been the case twenty years earlier (that is, in the 1890s):

> At Goalundo or Damukdia Stations, where one sees tons of fish being brought in, one can get a fish by begging, but never by purchase. There is no doubt that along the railway line and the steamer routes, the local markets are all being starved for the higher prices fetched in the distant markets. The growing influx of people into towns increases the demand and to meet the demand the interior of the country is robbed of its local supply.[59]

There were also allegations of local government neglect of fisheries in the region. In describing the causes of the famine of 1906, Sister Nivedita remarked that there was evidence of excessive taxation, including road and district taxes, the latter having doubled by the turn of the century. Nivedita's criticism was not directed against tax itself, but against the misuse of tax revenues, which were meant to be spent for the development of the respective districts. She lamented that a district board that was given a free hand in spending revenues for development works was dominated by the 'English conception of things', which involved the 'opening up' of the country by means of long roads and light iron bridges. Nivedita remarked that all such schemes of work 'fed the foreign engineers' and that none of them sought to restore native fisheries.[60] In the following decades, there were few actions taken on behalf of the district or union boards to contradict Nivedita's observations.[61]

From a nutritional point of view, fish is particularly important in a typical Bengali diet, which, apart from rice, mainly consisted of various pulses such as *Lathyrus sativa*, green gram, lentil and field-peas. The nutritional value of these pulses is relatively poor, but the inclusion of fish in the diet compensates for their nutritional deficiency.[62] At the same time, since fish eat mosquito larvae, a declining fish population meant an increase in the numbers of mosquitoes and a higher incidence of malaria.[63] Thus, the deterioration of fish stocks not only meant a reduced nutritional intake but encouraged an increase in malaria.

Nutrition was also compromised as a result of subtle changes in the availability of the aman variety of rice, which was considered nutritionally superior to other varieties. In 1937, a Bengali biologist observed that aman had a superior amino-acid make-up of protein to the aush variety and that it contained one or more 'growth essential factors' deficient or lacking in aush.[64] The protein value of both aman and aush varieties of rice was identical, the biological value being 80 in both cases, but in an experiment measuring the growth of young rats fed on both varieties of rice, aman was found to be far superior to aush. The biological value of aman rice, measured by the growth method, was 2.01, while the corresponding value for aush rice was 0 (zero), that is, the rats fed on aush rice just maintained their body weight.[65] As we have seen, the impact of the water hyacinth and changes to water regimes as a consequence of the construction of embankments, had forced people to replace aman with aush or boro varieties.[66]

Chunilal Bose's observation in the 1930s that the Bengalis received the least nutrition in the world clearly pointed to the decreasing consumption of ecologically embedded foodstuffs (Table 8.3). With a decreasing amount of nutrition from rice and fish, the bulk of the rural population might have been easy prey to starvation during the 1943–44 famine.[67] But in addition to looking at malnutrition and ecologically-induced diseases, one also needs to map the public entitlement to medicine and recovery. We may examine this through the case of malarial fever and cholera, two of the most commonly fatal diseases in colonial Bengal.

The colonial state seems to have been aware of the deteriorating conditions of public health, as is reflected in the documentation of various diseases and remedial efforts.[68] However, a curious link between the

Table 8.3   Comparative calorie intake

'Race'	Amount of protein available in daily diet (ounces)
Bhutanese	5–6
Tibetans	5
Punjabi	5
Sikh	4.5
Englishman	4
Japanese	3
Bengalis (east)	2.5
Bengalis (west)	2

Source: Chunilal Bose, *Food*, (Calcutta University, 1930), pp. 96, 101.

marketing of quinine and failure to ameliorate the ecological conditions that enabled malaria to flourish left anti-malaria efforts largely unsuccessful. It reflected contradictions between the state's programmes and its perception of links between environment and diseases and also in the way in which the human body was rendered a 'colony' of modern medicine.[69] In a government resolution of 1910, it was noted that in Eastern Bengal and Assam, annual fever-related deaths were estimated to be 700,000 and that about a quarter of these deaths were directly related to malaria. In this context, apart from stressing the need for further research, three major schemes were formulated: first, general improvement of physiological and sanitary conditions; second, extermination of mosquitoes; third, quinine treatment and prophylaxis. The resolution considered the first course of action to be concerned mostly with educating the people through government agencies. On the second issue, the resolution suggested that 'it would be most imprudent, especially in rural areas, to undertake anti-larval operations on any large scale until further enquiries have determined the precise ends to which labour and money can most profitably be directed'. In this context, the Lieutenant-Governor, Lancelot Hare, did not intend to do 'more than to continue the experiments which were being carried out at Dinajpur and Jalpaiguri'. The third line of action, however, was found to be 'more immediately practicable' and this option formed the subject of the resolution in question. Since there was 'no longer any doubt about the positive effect of quinine in the treatment of malaria', the resolution demanded that the use of quinine had to be 'more widespread, more regular and more effective'. According to the resolution, though the quantity of quinine annually sold in Eastern Bengal and Assam was quite large, constituting about three and four million doses respectively, it represented a very small amount compared to the population of the province and to the incidence of the disease. The resolution also referred to the ineffectual manner in which the drug was taken, that is, as a single dose during a malaria attack. The resolution therefore noted that this produced little or no effect on the disease which could be cured only by 'continuous and regular quinine treatment'.

Recognizing the tendency to 'haphazard' usage of the drug, the central feature of the proposed scheme appeared to be the 'distribution not of doses but of treatments, the unit of sale being a quantity of quinine sufficient to provide an effective remedy if properly used'. In the new formula, each course of treatment would consist of twenty 4-grain tablets packed in an airtight glass tube. Ten of these tubes were to make up a parcel and a parcel was to be the minimum amount sold to the retailer. The price of a parcel to the retailer was fixed at Rs 1.7, and as the price to the consumer was to be 3 annas per course, the retailer was supposed to earn a profit of 7 annas per parcel or more than 30 per cent on his outlay. It was observed that the profit was sufficiently high to stimulate sales automatically and so avoid the necessity of a paid agency for distribution. The resolution concluded that as treatment would be sold at less than cost price, the loss to government would be considerable. But if the project was successful and if the people were ready to accept the opportunity to alleviate their suffering, the Lieutenant-Governor felt that he would 'cheerfully face the loss', and he was convinced that provincial revenues could be 'devoted to no more worthy object'. Against the possible suggestion that the package would be too expensive to be accessed by the people, the Lieutenant-Governor argued that 'the peasantry of the province were well-to-do and an expenditure of 3 annas should not, except in the case of the very poorest, prove excessive'. He also suggested that it was better that the 'villager should spend 3 annas on a treatment which is likely to cure him than that he should throw away a *pice* on a remedy which can give him no more than passing relief'.[70]

This investment of provincial revenue in the marketing of quinine rather than in dealing with the sources of mosquito larvae was as much to do with an overwhelmingly rational conviction of the utility of quinine as with the intention to generate profit. Such orthodoxy about the power of modern medicine pushed the question of ecological issues further into the background.[71] Medicine's power to cure meant that nature was conquerable, hence there was less concern with its potential to create diseases. This attitude led the authorities to ignore, for example, the call to address the problem of the mosquito larvae that created havoc in areas choked with water hyacinth, the borrow pits of the embankments or stagnant water in the paddy fields.[72] For the same reason, the problem of larvae breeding in the Sundarbans clearance areas, which was the result of haphazard embankment, was also left unaddressed.[73] (see Figures 8.1 and 8.2.) The apparent failure of the government to appreciate the implications of deterioration of the deltaic water bodies for malaria was compounded by its failure to link it to the emerging global health organizations, such as that of the League of Nations. Even though a Malaria Commission was formed by the League, and in spite of the fact that malaria continued be the biggest threat to life and well-being in the Bengal Delta, throughout the 1930s the League's Malaria Commission focused solely on the problem of malaria in European deltas.[74]

As far as cholera was concerned, it appeared that prevention was indeed better than cure. Although *Comma bacillus*, the contributing agency of cholera, was only finally identified in 1884 after decades of research, the British physician John Snow had demonstrated as early as 1854 that contaminated water was largely responsible for the spread of the disease. As a consequence, efforts were made in England in the late nineteenth century to develop sewer systems and water treatment facilities. New technologies were invented to separate harmful waste from water. Chemical treatment, notably that of chlorination, killed bacteria in the water supply, making it safe to drink.[75] A determination to limit the transmission of cholera by providing safe drinking water resulted in two methods: one was the chemical method of cleaning the bacteria in the water supply by chlorination; the other relied on the deep tube well method. By the turn of the twentieth century, both methods were in use in India.

Measures to prevent the outbreak of cholera included systematic house-to-house visits and discussions between the people and Bengali sub-assistant surgeons. Each household was visited at least once during the year. Issues relating to cholera, including 'elementary principles of personal hygiene', were also discussed in public lectures delivered in weekly village markets and schools. These educational programmes also emphasized the importance of using the tube wells. Tube wells were of two kinds: the shallow tube well with a depth of 25 feet and less; and the deep tube-well, with a depth of more than 25 feet – founded when water could not be obtained at a shallower level. As far as the possibility of contamination of water was concerned, tube wells were universally accepted as the most reliable source of water. Traditional sources of water, such as 'pot wells',[76] tanks or ponds and the river, now came to be seen as unsafe sources of drinking water. T.H. Bishop, chief medical officer of the Hardinge Bridge project, argued that none of the three sources were reliable. He observed that 'pot wells' were open to contamination by dust, by vessels used for withdrawing water, and by articles (such as cloths) accidentally dropped into them. He thought that ponds were 'merely a large pool open to contamination from every possible source', and were used equally as a domestic water supply, for the washing of clothes, and for the watering of cattle. The same applied to the river water that was obtained 'almost invariably from back-waters, which were stagnant for the dry months, and subject to the same sources of pollution as the tank, but also, during the early part of the rains, received the accumulated shore filth by drainage, until the water level rises sufficiently to flush or to obliterate them'. Bishop referred to an experiment which revealed that 70 per cent of the household water supply collected from these traditional sources was 'grossly polluted'. While tube wells appeared to be the best possible option, Bishop, nonetheless, showed little interest in why cholera tended to be more common in regions adjacent to railways, where relatively more tube wells existed.[77]

While preventive measures such as the construction of tube wells were, in principle, intended to cover the needs of all regions and all people, it became clear over time that more attention was given to the regions where development works, such as the constructions of railways or bridges, had been taking place. Yet such development schemes not only eventually undermined the large-scale preventive measures; even within the individual 'sphere of operation' of such development projects disease prevention met with only moderate success. (Figure 8.3 shows the incidence of cholera in relation to the development of the railways in East Bengal.) Aspects of the attempts to prevent the spread of cholera around the site of the construction of the Hardinge Bridge in the Pabna district raise some interesting points regarding the evolving interaction between 'tradition' and 'modernization'. As the cholera prevention operation started in the region in October 1911, one of the difficulties faced by the engineers and doctors was, according to Bishop, 'racial and caste prejudice'. He observed that while in some villages the tube well was used by all, in others higher caste Hindus objected to its use by Muslims and lower caste Hindus. Another difficulty the prevention team encountered was the incidence of wilful damage of the tube wells by unknown people. Bishop suggested that it was 'common for village quarrels to take the form of injuring or polluting an opponent's well', and that it was probable that the lack of public access to such tube wells may have led to such quarrels.[78]

Bishop's notes on the failure of getting optimum benefit from tube wells point to the fact that while the tube wells were intended to substitute for indigenous patterns of water use, the modern techniques were not easily applicable to traditional habits. Invocation of caste or racial prejudice could not explain such failure. Underlying the problem was the wider issue of displacement of the very idea of the commons. The introduction of modern techniques such as tube wells not only failed to mitigate the problem of the deterioration of the commons due to high costs,[79] but also created tensions within society because people failed to attach a sense of ownership to tube wells as they did in the case of indigenous water sources. This is not to argue that traditional sources of water supply were harmless before the modernization drive began, but

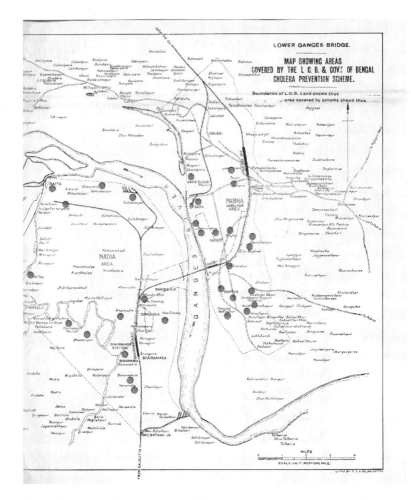

Figure 8.3   Map indicating the coexistence of railway construction and cholera
Source: T.H. Bishop, 'The Working of the Cholera Prevention Scheme', *IJMR*, 1(2) (1913–14).

the remarkable deterioration of the ecology in general and the upsurge
in diseases such as malaria and cholera largely coincided with the drive
for modernization. As Ira Klein has remarked, mortality in India during
the late nineteenth and early twentieth century 'showed one of the ways
that modernization interacted with indigenous conditions to produce
modernization crises, as traditional societies began to be transformed in
the image of the modern West'.[80] The modernization process continued
but sanitation and safe drinking water never became universal features

as a result of the state's misplaced appreciation of modernization process, budgetary constraints and the poverty of the rural population at large. The problems of public health therefore remained unabated in the years that followed – a fact that explains disease-related mortality during the great Bengal famine.[81]

9
Reflections

This book has shown how the patterns of agrarian dynamism and stagnation in colonial Bengal can only be understood by placing them in a long-term ecological perspective. Until recently, much of the historical scholarship on modern Bengal has dealt with a shorter time-frame, focusing either on the 'transition period' to colonialism or on the late colonial period leading up to decolonization. In pursuing a long-term historical perspective, this book has argued that the region's ecology made East Bengal a prosperous and dynamic part of South Asia's economy until far later than most historians imagine. This was a frontier region, drawing in capital, labour and intense imperial interest until at least the 1890s. Rural poverty in the area that is now Bangladesh emerged only in the twentieth century. As this book has argued, such downward shifts in agrarian economy and well-being stemmed from changes developing from the complex relationship between state, society and the region's highly fluid ecological regime.

Whilst examining changes in agrarian social relations and their impact on well-being, this book does not suggest that ecological history necessarily entails a study of unilateral environmental destruction. Against the tendency of some environmental historians to make a normative presumption in favour of conservation, this book has argued that the issue of 'conservation' versus 'destruction' is a more complex argument. Ecological regimes themselves create spaces that allow and sustain intervention; but the extent of threat and impact is determined by the perception of the community living in a particular ecological regime. The idea of ecology is alive only when the question of human involvement becomes prominent. The conservation ethos employed in the bulk of ecological studies accounts for the deterioration of the biosphere and biotic community, but it does not account for the social dynamics and processes which constitute the 'ecological'. Apologetic interpretations of colonial conservation practices and vehement criticism of ecological destruction during the colonial era equally fall into the prejudged format of pure conservation rhetoric without social considerations.

The nineteenth-century scenario in the Bengal Delta also contradicts to a certain extent the classical 'world system' theory's presumption that, in the context of oversupply of labour power at the periphery and coercive political conditions, workers are paid a decreasing amount per unit of production.[1] This study has shown that in deltaic Bengal, though labour supply was abundant, the colonial state and its intermediaries could not be fully coercive; neither were the workers underpaid. This was because, rather than being dictated by government policy, any form of ideology or the whims of the landlords, everyday life was dominated by the wilderness and fluid riverine atmosphere of the active delta, the agro-ecological potential of which could not be fully realized unless the labourers were given due share. Such development aided the growth of an agrarian society that created and used marketing facilities, dealt confidently with legal disputes with the landed elite and the state, and which developed an agrarian political process of its own, as reflected in the Faraizi movement. All these were possible because of an ecological regime that offered an expansive surface water network, optimum rainfall, extensive river transport and above all nature's own ambiguity, which obviated the coercive intrusion of the state's modern classificatory arms, as reflected in the decline of the Permanent Settlement of land and revenue governance.

The ecological contingencies that led to the decline of the Permanent Settlement were validated by the imperial scheme of a new province of 'Eastern Bengal and Assam' in 1905. By the 1880s, eastern Bengal was a major imperial centre for commodity production and international trade; rice and jute being the major products. About the same time the value of tea from Assam, which never came under the classic rules of the Permanent Settlement, was also established. For these two ecologically contiguous regions Chittagong port was assuming remarkable importance, particularly in the context of a lack of navigability at the port of Calcutta. It was not surprising that one of the reasons for the strength of the Swadeshi movement against the establishment of the new province, which was set to operate through the above dynamics, was the apprehension that it was the first step towards total abolition of the Permanent Settlement.[2] Yet, securing imperial 'success' in East Bengal and Assam through abolishing the Permanent Settlement in the era of 'high imperialism' was not the only consideration for the formation of the new province. In addition to the lure of the Indian Ocean, the possibility of connecting India with China through Burma also informed developments in this regard. Such a connection had underlain the demands of British commercial interests since the 1830s, particularly in the context of French commercial advances in South-east Asia.³ After the annexation of Burma, this demand gained additional momentum. Arthur Cotton proposed a water route, connecting the Brahmaputra and Yang-Tse rivers, but one Mc'Cosh thought that the route via Dhaka, Sylhet, Monipur, Ningtee or Kyen-duen river, upper Burma, Bhamo and Yunan appeared to have the 'advantage over every other route'. In fact, the first railway line in eastern Bengal in the 1860s between Calcutta and the central Bengal district of Nadia, was dubbed the 'first link between Calcutta and Canton'.⁴ By 1875, when Richard Temple was overwhelmed by the prosperity and commerce of eastern Bengal, it was discovered that in order to establish a connection between India and China via Burma, the Chittagong-Mandalaya route was the 'shortest and direct'.⁵ When in 1899, the Associated Chambers of Commerce, which was represented by 70 chambers of commerce in Britain, urged Curzon to connect India to China via Calcutta and Chittagong to Mandalay, it was merely echoing a demand that was almost a century old. Thus the formation of the new province of Eastern Bengal and Assam took place against the backdrop of its internal dynamics of agrarian mobility and its potential for trans-regional trade and commerce between India and China and across the Indian Ocean. The formation of the new province came at a time when, as data from the second year of its birth suggests, its total value of trade stood at just around 50 crore rupees, which was a sixth of the value of trade of the rest of India.⁶ A government report captured the industrial potential of the new province in these words:

> There is a general awakening of industrial feeling throughout the country, and that feeling has already taken practical shape in a few cases, and efforts are being made by the people themselves to launch into new forms of industrial enterprises. It needs, therefore, no powerful effort of the imagination to picture the day when, with her wonderful mineral and agricultural resources, and with her great natural advantages for both internal and foreign commerce, this province will surely occupy a prominent position in the industrial regeneration of this country.⁷

The nationalists who opposed the creation of the new province had a different take on the situation. They offered the nation a berth that was wide enough only for a 'Golden Bengal' centred on the paddy fields of the Bengal Delta. The revocation of the partition thus not only killed the possibility of breaking with what David Ludden described as 'spatial inequity'[8] across the British-Indian empire, but it also contributed to the speedy deterioration of agro-ecological resources through over-exploitation. The problem of agrarian stagnation by the turn of the twentieth century was closely linked to the ecological decline that had resulted from a variety of factors charted in this book: the extension of the railway embankments in a flat riverine landscape and the penetration of alien weeds, in particular the water hyacinth, were two such ecological forces. But while accounts of ecological deterioration tell much about the indifference, failure and mismanagement of the colonial state, agrarian decline also occurred as a consequence of the erosion of cultivators' entitlement to ecological resources. A major factor for such developments was the way in which the 'modern' bhadralok engaged with the agrarian world. Unemployment in urban centres, disease and poverty compelled the bhadralok to return to the smallholding peasant domain and its production processes. But the return of the bhadralok took place at a time when most fertile areas were already occupied and under cultivation and parts of the Sundarbans were cleared up to the sea.[9] The agro-ecological frontier that had continued to inform the dynamism of a distinct peasant society until the late nineteenth century was losing much of its formative space and pace. The bhadralok's renewed engagement with the actual agrarian production process within a squeezed agro-ecological resource base was associated with increased alienation of land from the actual cultivators and the steady draining of private capital from East Bengal to Calcutta.[10] The result was intense economic, social and political instability as reflected in the lack of well-being, extreme poverty and the spiral of communal violence that largely contributed to the partition of Bengal along with that of the subcontinent in 1947.

These findings and arguments raise several pertinent questions, particularly in relation to events and processes occurring during and after the partition. First, if the bhadralok invested and involved themselves so intensively in the East Bengal countryside, why were so many of them ready to jettison their gains by opting for a truncated Bengal, sacrificing also the possibility of collective political mobilization in line with linguistic nationalism? Partha Chatterjee suggests that the question can be examined with reference to the failure of Muslim political groups to engage with and ensure the security of the Hindu minority and to prioritize it in their emerging hegemonic polity. Chatterjee's reading of the problem is apposite, particularly in the context of the Noakhali communal riots of 1946, which made the Hindu minority more vulnerable than ever in East Bengal. In other words Chatterjee makes a strong case for the roots of Bengal's partition lying with the bhadralok's perception of insecurity. But Chatterjee, as well as other historians of communalism in late colonial Bengal, does not consider to what extent the severity of communal violence in Noakhali was connected to the fact that the district experienced the largest amount of land alienation in Bengal, possibly with the majority of land passing from cultivators to non-cultivating bhadralok. (see Table 5.2.) On the face of it, it appears that Chatterjee's approach to the question comes as a reaction to Joya Chatterji's highly politically focused argument that the bhadralok supported the partition because of the evident impossibility of acquiring higher berths in the power politics of Bengal, particularly in the context of the Communal Award of 1932.[11] On either count, it seems that the bhadralok's move in the fateful years leading to the partition was informed by a perceived sense of alienation and 'fear' of the majoritarian politics of Bengal's Muslims.

Not unexpectedly the political security attained by Bengal's bhadralok through partition was complicated by the anxiety over losing the ecological heart of the erstwhile 'Sonar Bangla'. This anxiety surfaced in the problematizing of the 'two nation theory' in the attempt to strike a political boundary that would leave Muslim majority Murshidabad in India and Hindu majority Jessore in East Pakistan, in order to secure a hydrologically suitable spot for diverting Ganges water.[12] The idea of taking water from the east-bound Ganges for West Bengal first surfaced in the early nineteenth century, but in the late colonial period the idea of tampering with the hydrological regime for development projects formed a major part of the modernizing programme of the nation. Meghnad Saha, for instance, envisioned a national project on the lines of the Tennessee Valley Authority for the Damodar valley in West Bengal and advocated similar multipurpose developments elsewhere along India's major rivers. In this context, Saha imagined 'a national purpose behind all planning'.[13] With the partition of Bengal a new and clearer form of ecological nationalism was programmed and articulated.[14] In 1948, Srischandra Nandy noted: 'our position as regards food production has been dangerously affected due to the major paddy-growing areas being made over to Eastern Bengal. In short, as a result of partition there is now the sad legacy of less food but more to be fed.' For Nandy, the slogan was: 'Produce or Perish!', a dilemma that was to be addressed by an improved river system. To him, the scheme of a barrage across the Ganges was perhaps the only effective means of inducing a 'copious supply of fresh up-land water' down through the Bhagirathi, the Hooghly and other dead and dying spill-channels in central Bengal. To him this was the surest way 'not only to put new life into our moribund countryside, but also to maintain the efficiency of the Port of Calcutta which has assumed a vital significance to us after the partition of Bengal.'[15] Over the years this anxiety to sustain West Bengal agriculture and economy culminated in the construction of the Farakka barrage, which was commissioned in 1975.

The artificial diversion of the Ganges was understandable in the context of both the precarious agro-ecological conditions of West Bengal and the political separation of the productive region of East Bengal. However, since its erection the Farakka barrage has been blamed for various ecological and humanitarian problems in West Bengal including serious erosion of land and soil, flooding and contamination by arsenic. For Bangladesh the barrage has continued to impact severely on fisheries, agriculture, cropping patterns, flooding, arsenic poisoning, erosion and so on, affecting about one-third of the total territory of Bangladesh with a population of about 50 million people.[16] Yet a larger version of this ecological nationalism is proposed through India's River-Linking Project, which aims to divert the waters of Ganges and Brahmaputra to the drought-prone areas of southern and northern India. There are concerns across South Asia that if implemented this project will multiply the ecological dislocations already caused by the Farakka barrage to such an extent that it will imperil not only Bangladesh, but the entire eastern Himalayan ecological regime, including the water system, and the Sundarbans forest region with all its biodiversity, as well as human and economic potential on a vast scale.[17] No wonder that South Asian environmental historians – and many others – are anxious about policymakers' inability to appreciate the formative historical developments that the natural hydrological system of the Ganges-Brahmaputra had promoted before the impact of capitalist modernization was fully felt in the region.[18]

An appreciation of the ecological imagination of the bhadralok and its contemporary implications provides a useful way of examining Bangladesh's agro-ecological predicament in the twenty-first century. An equally important question which emerges from the arguments in this book relates to the persistent problems of well-being in agrarian society, in particular in relation to ecological decline and political ecology inside the boundary of postcolonial Bangladesh. Since the famine of 1943, the food problem has never been fully resolved. Bangladesh continues to import food and starvation presents a continuing danger. Although there is general agreement that the 1980s and early 1990s saw increasing agricultural growth in Bangladesh, there is no consensus about the sustainability of this development.[19] Even if growth were there, it was not evenly observed across the country.[20] On the other hand, the process of land alienation has been so persistent that at present about half of the population remain functionally landless. Although maternal and child mortality have decreased to some extent, malnutrition and extreme poverty have remained a constant in the countryside since the late colonial period. All this despite the combined efforts of the state and non-government organizations and a large amount of foreign aid.

In relation to the problem of ecology itself, one cannot but see a remarkable continuity in the way that the postcolonial state has engaged the modernist paradigm of development efforts.[21] In addition to the existing 2900 kilometres of railway, since 1947 a huge number of embankments have been erected for surface transport; their emergence more the outcome of market logic than the result of any appreciation of ecological sustainability. The fascination with embankments has continued with their use in flood control schemes, often with unwanted consequences. For instance, the coastal embankments of the 1960s seem to have aggravated rather than mitigated the flood situation in Bangladesh.[22] Yet, constructing embankments rather than improving drainage remained high on the agenda for flood control, as was reflected in the controversial Flood Action Plan of the 1990s.[23] The plan has not been implemented, but it is alleged that many of its features crept into the National Water Policy promulgated in 1999. It is little surprise perhaps that in recent decades waterlogging has taken a much more dangerous turn than flooding itself. Much of the damage to crops resulting from waterlogging and pests can be explained as a consequence of such developments. The suspicion of surface water and love of deep tube-wells which emerged in the late colonial period has continued. But tube-wells have recently been blamed, along with other factors such as the Farakka barrage, for the serious problem of arsenic contamination to which more than a third of the population of Bangladesh as well as the people of West Bengal have been exposed.[24] The problem of the water hyacinth continues, although in a less severe form than in the past and also in the context of newer environmental problems. Still there are those in favour of complete eradication of the hyacinth, who refer to its links with cholera, malaria, dengue, depletion of fish resources and even climate change, and others who enthusiastically favour the utilization of the weed, for example, for making paper or toys, as a biofuel or for removing arsenic from water. The hyacinth has even been used to explain the cultural politics of feminism in Bangladesh by a feminist group who use the water hyacinth – a beautiful plant with attractive flowers, but at the same time a troublesome weed – as a metaphor for the peripheral condition of women in male-dominated society and, therefore, it 'challenges this concept as the women's movement does to the patriarchal notions'.[25]

If the colonial inheritance of ecological problems and the colonial state's engagement with nature powerfully inform many of today's sorry conditions of agrarian production and social economy, a far more important question relates to the forces of appropriation and exploitation. Apart from the problems of ecology itself there are issues relating to rural decline that could be explained by the way in which different social agencies appropriated limited ecological resources and rural 'surplus'. Overall, the well-being of the rural population is affected not only by static agrarian productivity but also – and more importantly – by threats to their entitlement to land, ecological resources and access to market. About half of the population is functionally landless at a time when the country has 3.3 million acres of khas mahals, of which 75 per cent are land and the rest water-bodies.[26] Although population growth has been a major factor in the lower average ratio of land per person, a redistribution of existing land would entitle each family to at least 0.5 acres. While this might not represent enough land to ensure a desirable standard of living, it could nonetheless reduce the number of landless people to a much lower level.[27] Why then is this not happening?

The political-ecological approach that this book has taken leads us to the conclusion that the problem of unequal access to ecological resources and economic underdevelopment can be understood by examining the different social forces that inform the particular power relations that sustain the process of de-entitlement. Of the total number of khas mahals, 88 per cent have been illegally occupied by the wealthier classes, whereas 60 per cent of people living in the chars are landless. Of the 1723 sq km of char lands, 93 per cent are occupied by only 23 per cent of the people. The lives of about 10.3 million people depend on river and water bodies. Of the existing 1.2 million acres of khas water bodies, only 5 per cent has been leased to the landless poor, 95 per cent being illegally occupied by more influential and richer segments of society who in turn develop unproductive patron-client relationships.[28] Like the historians of colonial times, contemporary observers, including anthropologists and economists, continue to examine the problem of rural decline by pointing at the patriarchal 'rich peasant' and his ethnic, communal and 'traditional' attributes.[29] But as we have argued in this book, the category of 'rich peasant' is a very fluid one in which the exploitative agency might remain hidden behind a semantic banner. The rich peasant only makes sense if seen in relation to the dynamics of the modern state.

Some recent works take a holistic approach to the 'agrarian' question and examine the problem of rural economic stagnation by going beyond the 'autonomous' world of the peasant. Khan observes that the painful process of land alienation might have led to the accumulation of land at the rich peasant's or the state's disposal and thereby opened the possibility of capitalist or at least proto-capitalist agricultural development. Alternatively, productive farming capacity could have been built at least in ecologically better regions. So why has this not happened in Bangladesh? This paradox resonates from the historical development of the late colonial period. One may seek to answer it by looking at the way in which agro-ecological resources and productive processes are being adversely affected by forces from outside the boundary of the 'agrarian', especially the unstable barometer of national power politics which stands against effective uses of such resources. The accumulation of rural surplus does not reflect any kind of logical upward mobility towards capitalist development; and this is not essentially because of the rich peasant's culturally-embedded inability to accelerate the process, but because the rise and fall of non-agrarian political forces negatively influence the process of productive accumulation.[30]

In the context of the longer-term political-ecological history of colonial and postcolonial Bangladesh, it appears that the task of telling the story of the state's adjustment to changing social power requires a deeper appreciation of the plurality of social forces that is often obscured by the discourse of entanglement between the colonial and the colonized or the elite and the subaltern or class and consciousness. Despite remarkable recent shifts towards appreciating the need to accommodate a pluralistic perception of agrarian relations, environmental histories of South Asia have so far largely focused on the 'scientific' rather than the 'social' index of the state's power. There are debates about the nature of the modernity created and invoked in the nationalist imagination, but there is little doubt that this modernity accepted the forms and institutions of the colonial state as key apparatuses to translate the agenda of the politically-charged nation. These 'modern' social forces controlled the language of the state's politics and bureaucracy and in the process entered and engaged the domain of the peasant, often resulting in the alienation of the latter from its own ecological and social space. If in the nineteenth century the colonial state and its interlocutors among the local elite were vulnerable to nature, by the early twentieth century, they had joined forces to make the most of it. Thus for one hundred years from the 1840s, the agency of the peasant was made to travel to the domain of the emerging modern nation now imagined and controlled by the 'bhadralok'. A key thrust of this book is an attempt to locate the unsettling role of ecology in the expression of social powers in the region; in other words, the way in which the nineteenth-century state's vulnerability to nature's 'difficult-to-know' realms was now compensated by the strengths of a new local social group who claimed to have modern knowledge of the agrarian environment that could inform the new modalities of state power. That process continues to inform governance policies and practices in the post-colonial era in Bangladesh and West Bengal in India.

Notes

1. Introduction

1. *'Diaras* and *Chars* often first appear as thin slivers of sand. On this is deposited layers of silt till a low bank is consolidated. Tamarisk bushes, a spiny grass, establish a foot-hold and accretions as soon as the river recedes in winter; the river flows being considerably seasonal. For several years the *Diara* and *Char* may be cultivable only in winter, till with a fresh flood either the level is raised above the normal flood level or the accretion is diluvated completely' (Haroun er Rashid, *Geography of Bangladesh* (Dhaka, 1991), p. 18).

2. For notes on geological processes of land formation and sedimentation in the Bengal delta, see W.W. Hunter, *Imperial Gazetteer of India*, vol. 4 (London, 1885), pp. 24–8; Radhakamal Mukerjee, *The Changing Face of Bengal: a Study in Riverine Economy* (Calcutta,1938), pp. 228–9; Colin D. Woodroffe, *Coasts: Form, Process and Evolution* (Cambridge, 2002), pp. 340, 351; Ashraf Uddin and Neil Lundberg, 'Cenozoic History of the Himalayan-Bengal System: Sand Composition in the Bengal Basin, Bangladesh', *Geological Society of America Bulletin*, 110 (4) (April 1998): 497–511; Liz Wilson and Brant Wilson, 'Welcome to the Himalayan Orogeny', http://www.geo.arizona.edu/geo5xx/geo527/Himalayas/, last accessed 17 December 2009.

3. Harry W. Blair, 'Local Government and Rural Development in the Bengal Sundarbans: an Enquiry in Managing Common Property Resources', *Agriculture and Human Values*, 7(2) (1990): 40.

4. Richard M. Eaton, *The Rise of Islam and the Bengal Frontier 1204–1760* (Berkeley and London, 1993), pp. 24–7.

5. For notes on the geological changes in the region, see J.E. Webster, *Eastern Bengal District Gazetteer: Noakhali* (Allahabad, 1911), p. 41; W.H. Arden Wood, 'Rivers and Man in the Indus-Ganges Alluvial Plain', *Scottish Geographical Magazine*, 40(1) (1924): 10; J. Seidensticker and A. Hai, *The Sundarbans Wildlife Management Plan: Conservation in the Bangladesh Coastal Zone* (Gland, 1983), p. 120; Sugata Bose, *Peasant Labour and Colonial Capital: Rural Bengal since 1770* (Cambridge, 1993), pp. 11–12.

6. Sidney Burrard, 'Movements of the Ground Level in Bengal', *Royal Engineers Journal*, XLVII (1933): 234.

7. Prestage Franklin, 'The Ganges and the Hooghly – How to Connect These Rivers by Converting the Matabanga into a Navigable Canal', in Sunil Sen Sharma (ed.), *Farakka – A Gordian Knot: Problems on Sharing Ganges Waters* (Calcutta, 1986), p. 20. (First published in 1861.)

8. S. Bhattacharya, 'Eastern India', in Dharma Kumar (ed.), *The Cambridge Economic History of India*, vol. II (Cambridge, 1982), pp. 270–1. For a general description of changes in the deltaic river system of Bengal and their influence on agricultural production, see Mukerjee, *The Changing Face of Bengal*.

9. Eaton, *The Rise of Islam*.

10. For two powerful manifestations of these debates, see Madhav Gadgil and Ramachandra Guha, *This Fissured Land: an Ecological History of India* (Delhi, 1992) and Richard H. Grove, *Green Imperialism: Colonial Expansion, Tropical Island Edens, and the Origins of Environmentalism, 1600–1860* (Cambridge, 1995). The debate was more recently rekindled in Gregory Barton's *Empire Forestry and the Origins of Environmentalism* (Cambridge, 2002).

11. The focus on forests and dams, wildlife and livestock or peripheral hills in environmental-historical work on South Asia is reflected in several recent works including Anu Jalais, 'The Sundarbans: Whose World Heritage Site?', *Conservation and Society*, 5(3) (2007): 1–8; Rohan D'Souza, *Drowned and Dammed: Colonial Capitalism, and Flood Control in Eastern India* (Delhi, 2006); Gunnel Cederlöf and K. Sivaramakrishnan (eds), *Ecological Nationalisms: Nature, Livelihoods, and Identities in South Asia* (Seattle, 2006); David Mosse, *The Rule of Water: Statecraft, Ecology and Collective Action in South Asia* (Oxford, 2003); Arun Agrawal and K. Sivaramakrishnans (eds), *Agrarian Environments* (Durham, NC and London, 2000); K. Sivaramakrishnan, *Modern Forests: Statemaking and Environmental Change in Colonial Eastern India* (Oxford, 1999); Willem van Schendel, 'The Invention of the Jummas: State Formation and Ethnicity in Southeastern Bengal', *Modern Asian Studies*, 26(1) (1992): 95–128.

12. Mahesh Rangarajan, *Fencing the Forest: Modernizing Nature. Forestry and Imperial Eco-Development 1800–1950* (New Delhi, 1999), p. 206.

13. Christopher V. Hill, *South Asia: an Environmental History* (Santa Barbara, 2008). Hill suggests: 'because India has always been primarily [a] peasant society, environmental histories, in deed if not in word, have been produced in South Asia for far longer than there has been an official discipline for them' (pp. xx).

14. Introduced in 1793, the Permanent Settlement was set to collect about £3 million annually from the zamindars of Bengal and Bihar. P.J. Marshall, *Bengal: the British Bridgehead, The New Cambridge History of India* (Cambridge: Cambridge University Press, 1988, reprinted 1990), p. 123.

15. For a major study of the ideological origins of the Permanent Settlement, see Ranajit Guha, *A Rule of Property for Bengal: an Essay on the Idea of Permanent Settlement* (Paris, 1963); for tenural, legal and economic aspects of the settlement, see S. Gopal, *The Permanent Settlement in Bengal and its Results* (London, 1948); Sirajul Islam, *Permanent Settlement in Bengal 1790–1819* (Dhaka, 1979); Badruddin Umar, *The Bengal Peasantry under the Permanent Settlement* (Dacca, 1972) (in Bengali); Akinobu Kawai, *Landlords and Imperial Rule: Change in Agrarian Bengal Society, c1885–1940*, 2 vols (Tokyo, 1986–1987). Works that trace the legacy of present-day rural underdevelopment to the Permanent Settlement include James Boyce, *Agrarian Impasse in Bengal: Institutional Constraints to Technological Change* (Oxford, 1987); F. Thomasson Jannuzi, *The Agrarian Structure of Bangladesh: an Impediment to Development* (Boulder, 1980).

16. As envisioned in the late 1960s by Eric Stokes, *The Peasant and the Raj: Studies in Agrarian Society and Peasant Rebellion in Colonial India* (New York, 1978).

17. Ratnalekha Ray, *Change in Bengal Agrarian Society* (New Delhi, 1979).

18. Rajat Ray, *Social Conflict and Political Unrest in Bengal 1875–1927* (New Delhi and Oxford, 1984), p. 51. See also Nariaki Nakazato, *Agrarian System of Eastern Bengal, c. 1870–1910* (Calcutta, 1994).

19. Bose's arguments, from the 1980s on, are summarized in Bose, *Peasant Labour and Colonial Capital*, pp. 84–90, 130–4, 162–9. For reinforcement of the view of the rich peasants role as rent-receivers, see Bidyut Chakrabarty, *The Partition of Bengal and Assam, 1932–1947: Contour of Freedom* (London and New York, 2004), pp. 36–41.

20. Sugata Bose, *Agrarian Bengal: Economy, Social Structure and Politics, 1919–1947* (Cambridge, 1986), pp. 162–4. See also Bose, *Peasant Labour and Colonial Capital*, p. 89. A similar argument is extended by Partha Chatterjee in *The Present History of West Bengal: Essays in Political Criticism* (Delhi, 1997), pp. 59–64 and in 'The Colonial State and Peasant Resistance in Bengal', *Past and Present*, 110(1) (1986): 182–9; Saugata Mukherji, 'Agrarian Class Formation in Modern Bengal 1931–1951', Occasional Paper no. 75, Centre for Studies in Social Sciences, Calcutta, 1985.

21. See Partha Chatterjee's review of Bose's *Agrarian Bengal*, in *Journal of Asian Studies* 47(3) (1988): 670–2.

22. Partha Chatterjee, 'Agrarian Relations and Communalism in Bengal, 1926–35', in Ranajit Guha (ed.), *Subaltern Studies 1: Writing on South Asian History and Society* (Delhi, 1982), p. 18.

23. A common term in the historiography of modern India, 'bhadralok' refers to a category in colonial Bengal that represents a hybrid constellation of upper caste, English-educated to illiterate, poorer to elite, salaried to rent-seeking, leftist to rightwing, world-class intellectual to mediocre, and orthodox imperialist to perennial nationalists. For debates on the definition and scope of the term, see M.N. Roy, 'Bourgeois Nationalism', *Vanguard*, 3(1) (1923); J.H. Broomfield, *Elite Conflict in a Plural Society: Twentieth-Century Bengal* (Berkeley, 1968), pp. 5–13; Gordon Johnson, 'Partition, Agitation and Congress: Bengal 1904–1908', *Modern Asian Studies*, 7(3) (1973): 534–5; Joya Chatterji, *Bengal Divided: Hindu Communalism and Partition, 1932–1947* (Cambridge, 1994), pp. 3–7.

24. For a book-length narrative of the East Bengal peasant's 'false consciousness', see Taj Hashmi, *Peasant Utopia: the Communalization of Class Politics in East Bengal, 1920–1947* (Boulder, 1992).

25. David Ludden, 'Introduction: a Brief History of Subalternity', in David Ludden (ed.), *Reading Subaltern Studies: Critical History, Contested Meaning, and the Globalization of South Asia* (Delhi and London, 2002), pp. 6–7.

26. Bose, *Agrarian Bengal*, p. 231.

27. Chatterjee, 'Agrarian Relations and Communalism in Bengal', p. 183.

28. David Ludden, *An Agrarian History of South Asia* (Cambridge, 1999), p. 2.

29. K Sivaramakrishnan, 'Situating the Subaltern: History and Anthropology in the *Subaltern Studies* Project', in Ludden, *Reading Subaltern Studies*, pp. 241–2.

30. Gyan Prakash, 'Subaltern Studies as Postcolonial Criticism', *American Historical Review*, 99 (December 1994): 1475–90. For a summary of the critique and the changes within the subalternist group regarding this question, see Ludden, *Reading Subaltern Studies*, pp. 13–20. See also Arild Engelsen Ruud, 'The Indian Hierarchy: Culture, Ideology and Consciousness in Bengali Village Politics', *Modern Asian Studies*, 33(3) (1999): 689–732.

31. See for instance, Gautam Bhadra, *Iman O Nishan: Unish shotoke bangaly krishak chaitanyer ek adhyay, c. 1800–1850* (Calcutta, 1994).

32. For a critique of the inflexibility of the understanding of 'agrarian' in the Indian context, see Agrawal and Sivaramakrishnan, *Agrarian Environments*, pp. 1–11.
33. For instance, James C. Scott, *Seeing Like a State: How Certain Schemes to Improve the Human Condition have Failed* (New Haven and London, 1998).
34. Grove, *Green Imperialism*.
35. C.A. Bayly, *Rulers, Townsmen and Bazaars: North Indian Society in the Age of British Expansion 1770–1870* (Cambridge, 1983).
36. Jon E. Wilson, *The Domination of the Strangers: Modern Governance in Eastern India* (Basingstoke, 2008).
37. Sivaramakrishnan, *Modern Forests*, pp. 30–66.
38. Gunnel Cederlöf, *Landscapes and the Law: Environmental Politics, Regional Histories, and Contests over Nature* (Ranikhet, New Delhi, 2008).
39. A.D.B. Gomess, Commissioner of the Sundarbans, in *Report on the Census of Bengal, 1871* (Calcutta, 1872), p. xi; it may be noted that Europeans were given the right to take leases of land in this region in 1829.
40. Bayly suggests that the long depression seems to have stimulated the structural changes that were already taking place in the Indian economy, paving the way for a 'levelling down of elite strata in Indian society and creation of a more homogeneous peasantry'. See C.A. Bayly, 'State and Economy in India over Seven Hundred Years', *Economic History Review*, n.s., 38(4) (1985): 591.
41. R.P. Tucker and J.F. Richards (eds), Editors' Introduction, *Global Deforestation and the Nineteenth-Century World Economy* (Durham, 1983), pp. xi–xv.
42. D.K. Fieldhouse, 'Colonialism: Economic', *International Encyclopedia of Social Sciences* (New York, 1968), p. 9.
43. See Michael Adas, *The Burma Delta: Economic Development and Social Change on an Asian Rice Frontier, 1852–1941* (Madison, 1974).
44. Bose, *Peasant Labour*, pp. 48–51; Muin-ud-Din Ahmad Khan, *History of the Fara'idi Movement in Bengal, 1818–1906* (Karachi, 1965); Narahari Kaviraj, *Wahabi and Farazi Rebels of Bengal* (New Delhi, 1982).
45. Scott, *Seeing Like a State*.
46. David Ludden, 'Agrarian Histories and Grassroots Development in South Asia', in Agrawal and Sivaramakrishnan, *Agrarian Environments*, p. 257.
47. C.A. Bentley, *The Times*, 13 October 1922, p. 11.
48. For a perceptive study of the governance of nature for 'development' in modern Britain and its imperial implications, see Richard Drayton, *Nature's Government. Science, Imperial Britain, and the 'Improvement' of the World* (New Haven and London, 2000).
49. Samir Amin, 'Globalism or Apartheid on a Global Scale?', in Immanuel Wallerstein (ed.), *The Modern World-System in the Longue Durée* (Boulder and London, 2004), p. 15.

2. Ecology and Agrarian Relations in the Nineteenth Century

1. Ranajit Guha, *A Rule of Property for Bengal: an Essay on the Idea of Permanent Settlement* (Paris, 1963); V.K. Gidwani, '"Waste" and the Permanent Settlement in Bengal', *Economic and Political Weekly*, 27(4) (1992): 39–46.

2. R.H. Hollingbery, *The Zemindary Settlement of Bengal*, vol. 1 (Delhi, 1985), p. 363 (first published Calcutta, 1879). Also note this comment: 'One of the principle causes of the change of opinion regarding the permanent settlement and of their order that no intermediate landed proprietors should be established between their European collectors and the ryots, is the desire to obtain all the revenue from the waste lands which may hereafter be brought into cultivation which is their declared intention "not to abandon"' ('Permanent Settlement of the Indian Land Revenue', *Asiatic Journal and Monthly Miscellany* (1929): 171).

3. James Mill, *The History of British India*, vol. 5 (London, 1826), pp. 545–6. See also Henry Leland Harrison, *The Bengal Embankment Manual Containing an Account of the Action of the Government in Dealing with Embankments and Water Courses since the Permanent Settlement: Discussion of the Principles of Act of 1873* (Calcutta, 1909), p. 18.

4. Sulekh Chandra Gupta, 'Retreat from Permanent Settlement and Shift Towards a New Land Revenue Policy', in Burton Stein (ed.), *The Making of Agrarian Policy in British India 1770–1900* (Delhi, 1992), p. 66.

5. 'Civis', 'The Affairs of India', *The Times*, 26 April 1842: 5.

6. Ibid.

7. See letter to Sir John Hobhouse, Bart, MP by 'Libra', 'State of India', *The Times*, 2 February 1839: 3.

8. *Rammohun Roy on the Judicial and Revenue System of India* (based on his answers to questions by the India Board), *The Times*, 1 August 1832: 6.

9. Hollingbery, *The Zemindary Settlement*, p. 362.

10. Henry Beveridge, *The District of Bakarganj: its History and Statistics* (London, 1876), p. 160.

11. F.E. Pargiter, *A Revenue History of the Sunderbans from 1765 to 1870* (Calcutta, 1885), p. 22; for a note on the zamindars' claim over the tracts of the Sundarbans that adjoined their permanently settled states and the government's negative response to such claims, see James Westland, *A Report on the District of Jessore: its Antiquities, its History, and its Commerce*, 2nd edn (Calcutta, 1874), pp. 107–10.

12. Pargiter, *A Revenue History of the Sunderbans*, p. 12; for a discussion of colonial revenue management in the Sundarbans, see John F. Richards and Elizabeth P. Flint, 'Long-term Transformations in the Sundarbans', *Agriculture and Human Values*, 7(2) (1990): 17–33.

13. J.B. Kindersley, *Final Report on the Survey and Settlement Operations in the District of Chittagong 1923–33* (Alipore, 1938), p. 49.

14. NAB, Offg Comm of the Chittagong div to Director of Agricultural dept, Bengal, no.733GC, 28 Nov 1885, in the Report by the Director of Agricultural dept on Management of Government and Ward's Estates and Creation of an Agriculture dept for the Lower Provinces, 1886, pp. 55–6.

15. F.H.B. Skrine, Offg Collector, Tipperah to W.W. Hunter, Director General of Statistics to GoI, 25 Jul 1885, in *Movements of the People and Land Reclamation Schemes* (Calcutta, 1885), p. 5.

16. H. Beveridge, *District of Bakarganj*, p. 182.

17. For a discussion on the process of land formation, see J.C. Jack, *Final Report on the Survey and Settlement Operation in the Bakarganj District, 1900–1908* (Calcutta, 1915), pp. 1–11, 109, 114. See also, Richards and Flint, 'Long-term Transformations in the Sundarbans', 17–33.

18. J.E. Webster, *EBDG: Tippera* (Allahabad, 1910), p. 86.

19. For instance, Panaullah, an ordinary cultivator, got a concession of Rs. 15 off the first instalment of rent for having brought to notice the existence of an island in Dhaka district. See NAB, W.G. Collection X, no. 34, Dacca Div, 'A' progs, Rev dept (Land), wooden bundle 16, list 17: H.H. Risley, Acting Assistant Secy, GoB to T.B. Lane, Secy, BoR, 12 Oct 1876.

20. Quoted in Ratan Lal Chakraborty and Haruo Noma (eds), 'Selected Records on Agriculture and Economy of Comilla District, 1782–1867', Joint Study on Agriculture and Rural Development (JSARD) Working Paper, no. 13 (Dhaka, 1989), p. 95.

21. *Act IX of 184: Alluvial Land, Bengal: Assessment of New Lands*; Beveridge, *The District of Bakarganj*, p. 181.

22. Henry Rickett, *Settlement Report of 1849*, p. 27.

23. IOR, V/24/2452–v/24/2504; an estate did not represent a fixed amount of land. For instance, a report mentioned about five government estates in Chittagong which comprised about 45,000 plots and 1,187,600 acres.

24. For instance, the decennial assessment of 1790, which was made permanently fixed in 1793, was very high and unequal in distribution. In the context of the total number of 14,500 zamindary estates of the Dhaka district, about one-third of the zamindars, paying about one-fourth of the total revenue demand, refused to accept the Permanent Settlement. The government brought these estates under its khas (direct) management. The same development took place in Tippera (Comilla) district. See Sirajul Islam, *Permanent Settlement in Bengal 1790–1819* (Dhaka, 1979), pp. 25–6. Sirajul Islam does not clarify what happened to these wastelands. But given the government's attitude towards the Permanent Settlement, it may be assumed that these were never returned to the zamindars.

25. NAB, Rev dept (Land), wooden bundle 47, list 17, file 16-S-4/1-2, no. 39–40, Feb 1902: F.A. Slacke, Secy to GoB, Rev dept, to Secy to GoI, no. 838 T, 24 June 1901.

26. NAB, Rev dept (Land), 'A' progs, wooden bundle 14, list 17: H.L. Dampier, Offg Secy to GoB to Offg Secy to BoR, Lower Provinces, 17 Aug 1868.

27. NAB, Rev dept (Land), 'A' progs, wooden bundle 14, list 17: H.L. Dampier, Offg Secy to GoB, to Offg Secy to BoR, Lower Provinces, 17 Aug 1868.

28. NAB, Rev dept (Land), 'A' progs, bundle 20, list 17, S.S. Collection XXI: H.J.S. Cotton to Secy to the GoB, 6 Aug 1881.

29. NAB, Rev dept (Land), 'A' progs, wooden bundle 14, list 17: H.L. Dampier, Offg Secy to the GoB to Offg Secy to BoR, Lower Provinces, 17 Aug 1868.

30. 'A Lover of Justice', *Permanent Settlement Imperilled or, Act X. of 1859 in its True Colors* (Calcutta, 1865), pp. 4–12.

31. IOR, P/66/26, Progs (Rev dept) of the Lieutenant-Governor of Bengal, 1860: Charles Wood, Secy of the State for India to Governor-General of India in Council, no. 5, 26 Jan 1860, p. 49.

32. Beveridge, *District of Bakarganj*, p. 161.

33. NAB, Rev dept (Land), 'A' progs, wooden bundle 14, list 17: C. Bernard, Offg Secy to GoB in Rev dept to Secy to Landholders and Commercial Association of British India, 18 Aug 1873.

34. F.D. Ascoli, *A Revenue History of the Sundarbans from 1870–1920* (Calcutta, 1921), p. 7.

35. B.R. Tomlinson, *The Economy of Modern India 1860–1970* (Cambridge, 1996), pp. 44–5.
36. James Rennell, 'Journal of Major James Rennell', T.H.D. La Touche (ed.), in *Bengal Asiatic Society Memoirs* III (1910–14); Matthew H. Edney, *Mapping an Empire: the Geographical Construction of British India, 1765–1843* (Chicago, 1997); Michael Mann, 'Mapping the Country: European Geography and the Cartographical Construction of India, 1760–90', *Science Technology & Society*, 8 (March 2003): 25–46.
37. Edney, *Mapping an Empire*, p. 340.
38. Westland, *District of Jessore*, p. 107.
39. IOR, P/903, Progs (Rev: Wastelands) of the Lieutenant-Governor of Bengal, Sept 1878, p. 17.
40. RAB, 1873–4, p. 34.
41. IOR, P/903, Progs (Rev: wastelands) of the Lieutenant-Governor of Bengal, Sept 1878, p. 17.
42. Secy to BoR, Lower Provinces, to Offg Secy to GoB, 21 Oct 1859, in *Papers Relating to Culturable Wastelands at the Disposal of Government* (Calcutta, 1860), p. 57.
43. W.W. Hunter, *A Statistical Account of Bengal*, vol.1 (London, 1875), p. 322; Westland remarked: 'So great is the evil fertility of the soil, the reclaimed land neglected for single year will present to the next year's cultivator a forest of reed (*nal*)' (Westland, *District of Jessore*, p. 178).
44. The story goes that when Henckell's agent was clearing parts of the Sundarbans, he was very much disturbed by tigers which attacked his people. He accordingly affixed to the place the name of Henckell, expecting that the tigers, dreading that name, would no more disturb him (Westland, *District of Jessore*, p. 106).
45. 'A Lover of Justice', *Permanent Settlement Imperilled*, p. 10.
46. J.W. Webster, *EBDG: Noakhali*, p. 82; For a detailed study of the Bengal land tenure sytem, see Sirajul Islam, *Bengal Land Tenure: the Origin and Growth of Intermediate Interests in the 19th Century* (Calcutta, 1988).
47. Beveridge, *District of Bakarganj*, p. 61. See also CSAS, Mukherjee Papers, Microfilm, article no. 8, box no. 13: 'Thirty nine articles on the Report of the Bengal Rent Law Commission' (reprinted from the *Hindu Patriot*, 1880–81), p. 8.
48. IOR, P/1488, Progs (Rev) of the Lieutenant Governor of Bengal, 21 Jan 1880, p. 2: Offg Secy to BoR to Secy to GoB.
49. IOR, P/1488, Progs (Rev) of the Lieutenant Governor of Bengal, no. 145, p. 3: Superintendent of Dearah Surveys, to Comm of Dacca, 8 Mar 1880.
50. NAB, Offg Comm of Chittagong Div to Director of Agricultural dept, Bengal, no. 733GC, 28 Nov 1885, in *Report by the Director of Agricultural Department on management of Government and Ward's Estates and creation of an Agriculture Department for the Lower Provinces*, 1886, p. 54.
51. IOR, P/1488, no. 145, Progs of Lieutenant Governor of Bengal, Rev dept, Feb 1880, p. 128: PC Roy, Superintendent of Dearah Survey, to Comm of Dacca, 8 Mar 1880.
52. Sirajul Islam, *Bengal Land Tenure*.
53. 'Takavi advances are a *sine qua non* to agricultural settlements in this country', see F.H.B. Skrine, Offg Collector, Tipperah to W.W. Hunter, 25 July 1885, in *Movements of the People*, pp. 5–6.

54. Hunter, *Statistical Account*, vol. 1, pp. 338–9.
55. Westland, *District of Jessore*, p. 178.
56. MoE in RIC (Calcutta, 1860), answer no. 2438; Also note the remark of a contemporary observer: 'There are no ryots in India, or in Britain either, who are more considerately treated, or who have more encouragement to better their condition, than the tenants upon the extensive zemindaris owned by the Messrs Morrell in Eastern Bengal' ('The Bengal Ryot', *Blackwood's Edinburgh Magazine*, cxiii, dclxxxviii (February, 1873): 161).
57. NAB, Rev (Land), 'A' progs, wooden bundle 17, no. 296A: D.J. McNeile, Offg Secy to BoR, Lower Provinces to Secy GoB, Rev dept, 6 June 1873.
58. IOR-P/66/26, *Rules for the Grant of Wasteland in the Soonderbuns*, 24 September 1853, Progs (Rev dept) of the Lieutenant Governor of Bengal, Mar 1860, pp. 138–9.
59. IOR, E/4/854, India and Bengal Despatches (Public Works), 4 Aug 1858, no.115, pp. 820–42.
60. F. Ascoli, *A Revenue History of the Sunderbans*, pp. 7–8.
61. NAB, Rev dept (Land), 'A' progs, wooden bundle 18, list 17: H.L. Dampier on the subject of 'Protection of Ryots who have not acquired rights of occupancy against enhancement of rent during the currency of a temporary settlement – arising out the Chittagong Noabad Settlement.'
62. Quoted in 'A Lover of Justice', *Permanent Settlement Imperilled*, p. 13.
63. West Bengal State Archives (WBSA), Rev P.V. June 1872-166: H.L. Dampier, Secy to GoB, Rev dept to Offg Secy, to BoR, no. 2337, 14 June 1872.
64. WBSA, Sept 1872-169, p.94: A.O. Hume, Secy to GoI, Agriculture dept, to Secy to GoB, Rev dept, no. 720, 8 Aug 1872.
65. These views of the Bengal Government are heavily drawn from WBSA, no. 1P, 27 Aug 1872, pp. 96–9: Offg Secy to GoB, Rev dept to Secy to GoB, Agriculture dept.
66. See George Campbell, *Memoirs of my Indian Career*, vol. II (London, 1893), pp. 215, 295.
67. 'A Lover of Justice', *Permanent Settlement Imperilled*, pp. 8–9.
68. NAB, Rev dept (Land), 'A' progs, wooden bundle 15, list 17, MIS – collection III, no.11: Khawja Ahsunollah and other Zemindars of Eastern Bengal, to Richard Temple, 1875.
69. NAB, Rev dept (Land), 'A' progs, wooden bundle 15, list 17, MIS – collection III, no.11: C.E. Buckland, Offg Junior Sec to the GoB, Rev dept to Khawja Ahsunoollah, and other zamindars of Eastern Bengal, 30 June 1875.
70. A.W. Paul, offg Collecor of the 24-Pergunnahs, to Hunter, 8 Sept 1885, in Hunter, *Movements of People*, p. 21.
71. Hunter, *Statistical Account*, vol. VI, p. 309.
72. A. Manson, Collector of Chittagong to Hunter, in *Movements of People*, p. 14. For an important narrative of similar circumstances by the turn of the nineteenth century, see Jon E. Wilson, ' "A Thousand Countries to go to": Peasants and Rulers in Late Eighteenth-Century Bengal', *Past & Present*, 189 (2005): 81–109.
73. IOR, V/27/314/23: P.M. Basu, *Survey and Settlement of the Dakhin Shahbazpur Estates in the District of Backergunge, 1889–1895* (Calcutta, 1896).
74. WBSA, Sept 1872-169: Offg Secy to GoB, Rev dept to Secy to GoB, Agriculture dept, no. 1P, 27 Aug 1872.

75. WBSA, Rev P.V. June 1872-166: H.L. Dampier, Secy to the GoB, Rev dept to Offg Secy, to BoR, Land Rev dept, letter no. 2337, 14 June 1872.
76. *Bengal Times*, 22 March 1876: 190.
77. IOR, Collector of Chittagong, to Secretary of Board of Revenue, No. 94, 10 March 1826, Progs (Revenue) of the Government of Bengal, March 1877, p. 40.
78. Webster, *EBDG: Noakhali*, p. 81.
79. A Manson, coll of Chittagong, to Hunter, 7 Aug 1885, in Hunter, *Movements of the People*, p. 13.
80. NAB, Offg Commissioner of the Chittagong Division to Director of the Agricultural dept, Bengal, no. 733GC, 28 November 1885, in Report by the Director of Agricultural dept on management of Government and Ward's Estates and Creation of an Agriculture dept for the Lower Provinces, 1886, p. 53.
81. M. Finucane, Director of the Agriculture dept, Bengal, to Secretary to the BoR, Land Revenue dept, in Report by the Director of Agricultural dept on management of Government and Ward's Estates and Creation of an Agriculture dept for the Lower Provinces, 7 July 1886, p. 19.
82. WBSA, F.R. Cockrell, Commissioner of Dacca to Secretary to the Board of Revenue, Lower Provinces, camp: Faridpur, 25 November 1874, Revenue R.V. January 1874-197, p. 4; for the political consideration about soft attitude to rent; see also: Para 9: 'The Lieutenant-Governor has found that some zealous officers in charge of wards' estates, who fancy that the only test of good management is to show an increased rent-roll, have undertaken a crusade against the rent-free holders and have thereby brought on Governor all the odium of such a measure [p. 97]', Offg Secretary to the GoB, Revenue dept to the Secy to the Government of India, Agriculture dept, no. 1P, 27 August 1872, in WBSA, September 1872-169, pp. 96–9.
83. Sugata Bose, *Peasant Capital*, p. 89.
84. Sirajul Islam, *Rent and Raiyat: Society and Economy of Eastern Bengal, 1859–1928* (Dhaka, 1989); Sugata Bose, *Peasant Labour and Colonial Capital. Rural Bengal since 1770* (Cambridge, 1993), p. 120.

3. Economy and Society: the Myth and Reality of 'Sonar Bangla'

1. The term 'Sonar Bangla' came into wide use in the early twentieth century on the wave of nationalist imagination of Bengal's prosperous past. However such reference to a golden Bengal was always more a strategy for political mobilization than a description of historical reality.
2. A recent major report on the global environment uses the term 'eco-system services' to refer to those 'benefits people obtain from ecosystems' which include '*provisioning services* such as food, water, timber, and fiber; *regulating services* that affect climate, floods, disease, wastes, and water quality; *cultural services* that provide recreational, aesthetic, and spiritual benefits; and *supporting services* such as soil formation, photosynthesis, and nutrient cycling'. See Preface to the Millennium Ecosystem Assessment, *Ecosystems and Human Well-being: Synthesis* (Washington DC, 2005).

3. For a description of geo-morphological influences on East Bengal's land-scape, soil fertility and fluvial system, particularly in comparison to Western Bengal, the Indus Valley and the Punjab, see Arthur Geddes, 'Alluvial Morphology of the Indo-Gangetic Plain', *Transactions and Papers* (Institute of British Geographers), 28 (1960): 253–76. For an overview of the effects of changes in the river system on salinity, see H. Viles and T. Spencer, *Coastal Problems: Geomorphology, Ecology and Society at the Coast* (London, 1995).

4. Joseph Dalton Hooker, *Himalayan Journals* (London, 1854), p. 339.

5. W.S. Seton-Karr, 'Agriculture in Lower Bengal', *Journal of the Society of Arts*, 16 (March 1883): 419.

6. W.W. Hunter, *A Statistical Account of Bengal*, 2 vols (London, 1875), vol. I, p. 288.

7. For a narrative of the interplay of the economy at better and worse periods during the second half of the eighteenth century, see Rajat Datta, *Society, Economy and the Market. Commercialization in Rural Bengal c.1760–1800* (New Delhi, 2000), esp. pp. 238–71.

8. G.E. Gastrell, *Geographical and Statistical Report of the Districts of Jessore, Fureedpore and Backergunge* (Calcutta, 1868), p. 6.

9. Seton-Karr, 'Agriculture', p. 421; some observers suggested that it could grow twelve inches in 24 hours and survive inundation for seven to eight days if the water was clear. See *Calcutta Review* (January 1874): 171.

10. Seton-Karr, 'Agriculture', p. 422.

11. '...[in the Sundarbans] you look over one vast plain, stretching for miles upon each side, laded with golden grain: a homestead is dotted about here and there, and the course of the rivers is traced by the fringes of low brushwood that grow upon their banks; but with these exceptions one sees in many places one unbroken sea of waving dhan [paddy], up to the point where the distant forest bounds the horizon' (James Westland, *A Report on the District of Jessore: its Antiquities, its History, and its Commerce*, 2nd edn (Calcutta, 1874), p. 182).

12. Gordon T. Stewart, *Jute and Empire: the Calcutta Jute Wallahs and the Landscapes of Empire* (Manchester, 1998), p. 38.

13. Hunter, *Statistical Account*, vol. II, p. 66.

14. Sugata Bose, *Peasant Labour and Colonial Capital. Rural Bengal Since 1770* (Cambridge, 1993), p. 53; for a statistics of spectacular growth of trade in jute between 1828 and 1873, see Hem Chunder Kerr, *Report on the Cultivation of, and Trade in, Jute in Bengal*, (Calcutta, 1877), p. 71.

15. D.R. Lyall, Comm of Chittagong Div, to Secy to GoB, 28 Apr 1888, in the *Report on the Condition of the Lower Classes of Population in Bengal* (Calcutta, 1888).

16. The Calcutta and Eastern canals and Tolley's Nullah constituted the major natural-cum-artificial navigation system. By the 1850s the system not only catered to 'the whole trade of eastern districts which was daily growing in importance', but in the dry season it also saw many trading boats from the Northwest Frontier provinces, which were unable to sail at this time through the western Bengal rivers. According to one estimate, this network of canals facilitated the traffic of about a million tons of goods per annum in the first decade of the twentieth century. The value of these goods was estimated at nearly four million pounds sterling. See W. Hunter ed., *Imperial Gazetteer*

of India, vol. IX (London: Trübner & Co., 1881–85), p. 287. These figures are also quoted in James Rennell, *Journal of Major James Rennell*, ed. T.H.D. La Touche, in *Bengal Asiatic Society Memoirs* III (1910–14), footnote by the editor, p. 20.

17. H.J.S. Cotton, 'The Rice Trade in Bengal', *Calcutta Review*, CXV (January, 1874): 174–8. For an important account of the external and internal factors and patterns of commercialization of agriculture in Bengal, see Binay Bhushan Chaudhuri, 'Growth of Commercial Agriculture in Bengal – 1859–1885', *Indian Economic and Social History Review*, 7(1) (1970), 25–60.

18. C. Addams Williams, *History of the Rivers in the Gangetic Delta 1750–1918* (Calcutta, 1919, reprinted Dacca, 1966).

19. Lal Behari Day, *Bengal Peasant Life* (London, 1892), pp. 169–70.

20. Instituted in 1861 to enquire into the anti-indigo movement.

21. Minutes of Evidence (MoE), answer no. 2413, in *Report of the Indigo Commission* (RIC).

22. Westland, *District of Jessore*, p. 184.

23. While crossing through the Sundarbans in the mid-nineteenth century, Hooker was surprised to see his boat being overtaken by a steamer in a creek, which in some places seemed hardly broad enough for it to pass through. He described this as 'a novel sight' and a strange experience. See Hooker, *Himalayan Journals*, p. 354.

24. This is for the year 1882–83. See W.W. Hunter, *The Indian Empire: its People, History, and Products* (London, 1893), p. 48.

25. Richard Temple, *Men and Events of my Time in India* (London, 1882), pp. 417–18.

26. Henry Beveridge, *The District of Bakarganj: its History and Statistics* (London, 1876), p. 284.

27. Abdur Rahim, *Akaler Puthi* [Songs of Famine] (Mymensingh, 1875), pp. 13–14.

28. *Bengal Times*, 12 February 1876, pp. 101, 108.

29. Westland, *District of Jessore*, p. 111.

30. Binay Bhushan Chaudhuri, 'Growth of Commercial Agriculture in Bengal': 30–1.

31. S.A. Latif, *Economic Aspect of the Indian Rice Export Trade* (Calcutta: Dasgupta & Co., 1923), p. 15.

32. Ibid., pp. 20–1.

33. Rajat Datta, *Society, Economy, and the Market*; Tilottama Mukherjee, 'Markets, Transport and the State in the Bengal Economy, c.1750–1800', unpublished PhD thesis, University of Cambridge, 2004; Sayako Miki, 'Merchants, Markets and the Monopoly of the East India Company: the Salt Trade in Bengal under Colonial Control, c.1790–1836', unpublished PhD thesis, SOAS, University of London, 2005.

34. Bose, *Peasant Labour*, pp. 84–9.

35. Willem van Schendel and Aminul Haque Faraizi, *Rural Labourers in Bengal, 1880 to 1980* (Rotterdam, 1984).

36. *Report on the Condition of the Lower Classes* (also known as the Dufferin Report). For a summary of the report, specific to the conditions of the labourers, see Willem van Schendel, 'Economy of Working Classes', in Sirajul Islam (ed.), *History of Bangladesh*, vol 2 (Dhaka, 1997), pp. 637–43.

37. P. Nolan, Secy to GoB, to Secy to GoI, 30 June 1888; *Condition of the Lower Classes*, p. 2.
38. Anundoram Borooah, Offg Collector of Noakhali to Comm of Chittagong Div,14 Apr 1888, in *Condition of the Lower Classes*.
39. D.R. Lyall, to Secy to GoB, Rev dept, 28 Apr 1888, in *Condition of the Lower Classes*.
40. W.R.Larmine, Comm of Dacca Div to Secy to GoB, Rev dept, 18 May 1888, in *Condition of the Lower Classes*.
41. This is a fairly liberal estimate. Raja Ram Mohun Roy found the requirement to be between 1 and 1.5 lb, which was about the same amount. See Raja Ram Mohun Roy, *Exposition of the Practical Operation of the Judicial and Revenue Systems of India* (London,1832), p. 106.
42. H. Savage, Offg Collector of Bakarganj to Comm of Dacca Div, 20 Apr 1888, in *Condition of the Lower Classes*.
43. Baboo Haripada Ghose, Tahsildar, Town Khasmahal, Chittagong to Magistrate of Chittagong, 8 May 1888, in *Condition of the Lower Classes*.
44. Haripada Ghose to Magistrate of Chittagong, 8 May 1888, in *Condition of the Lower Classes*.
45. J.C. Jack, *Final Report on the Survey and Settlement Operation in the Bakarganj District, 1900–1908* (Calcutta, 1915), p. 45. See also Eaton, 'Human Settlement and Colonization in the Sundarbans', *Agriculture and Human Values*, 7(2) (1990): 12.
46. Lyall to Secy to GoB, 28 Apr 1888; Nolan to Secy to GoB, 30 June 1888; Larminie to Secy to GoB, 18 May 1888, in *Condition of the Lower Classes*.
47. It may be noted that although the government periodically attempted to restrict the fishermen's traditional right to fish in riparian commons such as beels and rivers, it was not able to do so until the 1880s or in some cases until the end of the century. See Bob Pokrant, Peter Reeves and John McGuire, 'Bengal Fishers and Fisheries: a Historiographical Essay', in Shekhar Bandyopadhyay (ed.), *Bengal: Rethinking History. Essays in Historiography* (New Delhi, 2001), pp. 96–101.
48. Larminie to Secy to GoB, 18 May 1888, in *Condition of the Lower Classes*.
49. H. Savage, Offg Collector of Bakarganj to Comm of Dacca Div, 20 Apr 1888, in *Condition of the Lower Classes*.
50. For a useful summary of the income, mobility and living conditions of labourers in Bengal as cited in the Dufferin Report, see Schendel and Haque Faraizi, *Rural Labourers*, pp. 9–23.
51. Munshi Nandji, Settlement Officer, Nulchira Estate to Collector of Noakhali, 3 Apr 1888, in *Condition of the Lower Classes*.
52. Nolan to Secy to GoB, 30 June 1888, in *Condition of the Lower Classes*.
53. Schendel and Haque Faraizi, *Rural Labourers*, p. 10.
54. Settlement Deputy Collector to Jawar Baluakandi, Tippera to Director of dept of Land Records ad Agriculture, 24 Mar 1888, in *Condition of the Lower Classes*.
55. Lyall to Secy to GoB, 28 Apr 1888, in *Condition of the Lower Classes*.
56. P. Nolan to Secy to GoB, 30 June 1888; Larminie to Secy to GoB, 18 May 1888, in *Condition of the Lower Classes*.
57. Quoted in Schendel and Haque Faraizi, *Rural Labourers*, p. 20.
58. A.C. Tute, offg collector of Pabna, to W.W. Hunter, *Movements of the People and Land Reclamation Schemes* (Calcutta, 1885), p. 11, 5 August 1885, p. 12.

59. Quoted in Lyall to Secy to GoB, 28 Apr 1888, in *Condition of the Lower Classes*.
60. Schendel and Haque Faraizi, *Rural Labourers*, p. 10; the authors identify the rate as follows: for East Bengal districts, that is Mymensingh (4%), Tippera (6%) and Noakhali (6%), Chittagong (23%); for West Bengal Districts, Midnapur (43%), Malda (42%) and Murshidabad (37%). It is important to note that the highest number of labourers were found in Chittagong where the condition of the labourers were best in all East Bengal.
61. Nolan to Secy of GoB, 30 June 1888, in *Condition of the Lower Classes*.
62. The reverse was the case in East Bengal where the middle class or the bhadralok was relatively worse off, Nolan to Secy to GoB, 30 June 1888, in *Condition of the Lower Classes*.
63. Ibid., pp. 3–6.
64. 'Minute by the Lieutenant Governor of Bengal', 14 Jan 1876, cited in the RAB, 1874–75, p. 42.
65. Evidence of M.L. Hare, Comm of Dacca, to Indian Famine Commission 1898, *Report of the Indian Famine Commission, 1898*, appendix vol. 1: Evidence of witness, Bengal (Calcutta, 1898), p. 124.
66. William J. Collins, 'Labour Mobility, Market Integration, and Wage Convergence in Late 19th Century India', *Explorations in Economic History*, 36 (3) (1999): 246–77.
67. Tirthankar Roy, 'Globalization, Factor Prices and Poverty in Colonial India', *Australian Economic Review*, 47(1) (2007): 81.
68. APAC, MSS Eur F86/161, Temple Collections, 'Conditions of Peasantry', p. 1: C.E. Buckland to all Commissioners, August 1875; E.E. Lowis, Comm of Chittagong Div to Buckland, Private Secy to Lieutenant-Governor of Bengal, 16 Sept 1875, p. 2.
69. For notes on indigo cultivation and peasant resistance against it, see the next chapter.
70. APAC, MSS Eur F86/161, Temple Collections: J.R. Cockerall, Deputy Comm of Darjeeling, to Buckland.
71. APAC, MSS Eur F86/161, Temple Collections: F.B. Peacock, Comm of Dacca to Buckland, 4 Nov 1875, pp. 8–10.
72. Richard Temple, quoted in the *Bengal Times*, 29 January 1876: 68.
73. Collins, 'Labour Mobility, Market Integration', p. 270.
74. R.C. Dutta, *The Economic History of India* in *the Victorian Age*, vol. II (London, 1904), pp. vi–vii; also quoted in Bose, *Peasant Labour*, pp. 120–1.
75. William Twining, *Clinical Illustrations of the More Important Diseases of Bengal*, vol. I (2nd edn, Calcutta, 1835), p. 6.
76. Ibid., pp. 8–9.
77. John M'Clelland, *Sketch of the Medical Topography or Climate and Soils, of Bengal and the N.W. Provinces* (London,1859), pp. 42, 47–8.
78. Twining, *Clinical Illustrations*, p. 6.
79. Kazi Ihtesham, 'Malaria in Bengal from 1860 to 1920: a Historical Study in a Colonial Setting', unpublished PhD thesis, University of Michigan, 1986, p. 239. Also note this comment: 'Barring a few districts, Eastern Bengal as a whole enjoyed advantages in term of health and economic prosperity for which it would be difficult to find a parallel in any part of India.' Quoted in Arabinda Samanta, *Malarial Fever in Colonial Bengal 1820–1939: Social History of an Epidemic* (Kolkata, 2002), p. 23.

80. Twining, *Clinical Illustrations*, pp. 31–2.
81. Kanny Loll Dey, *Hindu Social Laws and Habits Viewed in Relation to Health* (Calcutta, 1866), pp. 3–5.
82. Baboo Isser Chunder Mitter Roy, 'A Few Facts Concerning Village Life', The Bengal Social Science Association Meeting, 1877 (Calcutta, 1877), p. 3.
83. Mitter Roy, 'A Few Facts Concerning Village Life', p. 3.
84. M'Clelland, *Sketch of the Medical Topography*, p. 29.
85. Ibid., p. 32.
86. Ibid., p. 38; M'Clelland noted that in response to the threat of such diseases, the Bengalis showed 'instinctive objection to live on ground floors. Their houses were consequently raised on posts. This practice, which was also prevalent in parts of Assam and Pegu, seemed to have originated in a perception of the capillary attraction of the soil, in consequence of which, the surface was always wet. Thus the natives of these regions displayed a just appreciation of the influence of soil', which became 'engrafted even in their national character and customs' (ibid., p. 126).
87. Census 1901, vol. VI, part I, 1902, p. 75.
88. Census 1901, vol. VI, part 1, 1902, Faridpur Dist, part 1, p. 77.
89. Larminie, Comm of Dacca, to Secy to GoB, 18 May 1888, in *Condition of the Lower Classes*.
90. Excerpts from paragraphs 37 to 39 from the General Administration Report, Chittagong Div for 1881–82, in *Condition of the Lower Classes*.
91. A. Borooah, to Comm of Chittagong Div, 14 Apr 1888, in *Condition of the Lower Classes*.
92. The Settlement Deputy Collector's notes, 24 March 1888, in *Condition of the Lower Classes*.
93. NAB, progs 'A', Rev (Land), wooden bundle 6, list 17: C.A. Kelly, Offg Collector, Faridpur to Comm of Dacca Div, 18 July 1870.
94. Hunter, *Statistical Account*, vol.1, pp. 160–1.
95. *Principal Heads of the History and Statistics of the Dacca Division* (Calcutta, 1868), p. 139.
96. Lyall to Secy to GoB, 28 Apr 1888, in *Condition of the Lower Classes*.
97. Ibid.
98. H. Sasson, Offg Collector, Bakarganj to W.W. Hunter, 7 Sept 1885, in *Movements of the People*, p. 24.
99. CSAS, Mukherjee papers: 'Thirty nine articles on the Report of the Bengal Rent Law Commission', p. 8.
100. APAC, Temple Collection, MSS Eur F86/165, misc. colln. 14, no. 26/27J.G: Offg Joint Magistrate of Munshiganj to Collector of Dacca, 19 Sept 1873; Seton-Karr, 'Agriculture', p. 425; see also MoE, answer no. 458.
101. Beveridge, *District of Bakarganj*, p. 191.
102. Sumit Guha, 'Agrarian Bengal, 1850–1947: Issues and Problems', *Studies in History*, n.s., 11(1)(1995): 125; see also James Alexander, 'On the Tenures and Fiscal Relations of the Owners, and Occupants of the Soil in Bengal, Behar, and Orissa', *Journal of Asiatic Society of Bengal*, XIV (July–December, 1845): 536–7.
103. *Condition of the Lower Classes*, p. 11.
104. F.H.B. Skrine, Offg Collector, Tipperah to W.W. Hunter, 25 July 1885, in *Movements of the People*, p. 7.

105. *Movements of the People*, p. 12.
106. Sekhar Bandyopadhyay, *Caste, Protest and Identity in Colonial India: the Namasudras of Bengal, 1872–1947* (London, 1997), p. 22.
107. For a comprehensive account of the emergence and rise of the Namasudras in eastern Bengal, see ibid., pp. 15–29; see also R. Carstairs, *The Little World of an Indian District Officer* (London, 1912), p. 58; For a note on the background of the rise of the non-elite tilling groups in India in the nineteenth century, see Susan Bayly, *Caste, Society and Politics in India from the Eighteenth Century to the Modern Age* (Cambridge, 1999), pp. 108–10, 200–2.
108. Richard M. Eaton, *The Rise of Islam and the Bengal Frontier 1204–1760* (Berkeley and London, 1993).
109. *Ghazi kalu and champaboti kanyar puthi* (Mymensingh, 1870).
110. Siddik Ali, *Siddik Alir Puthi*, 1244 BS (1837 AD).
111. J.E. Webster, *Eastern Bengal District Gazetteer: Noakhali* (Allahabad, 1911), p. 279.
112. Muzaffar Ahmad, *Myself and the Communist Party of India* (Calcutta, 1970), pp. 1–5.
113. Abdur Rahim, *Akaler Puthi*. (Author's translation.)
114. *Aftab-i-hidayat* (Mymensingh, 1877).
115. Asim Roy, *The Islamic Syncretistic Tradition in Bengal* (Princeton, 1983); Rafiuddin Ahmed, *The Bengal Muslims 1871–1906: a Quest for Identity* (Delhi, 1981).
116. Binay Bhushan Chaudhuri, 'Growth of Commerical Agriculture in Bengal'.
117. Shahid Amin, 'Dimensions of Dependence', in Ludden (ed.), *Agricultural Production and Indian History* (Oxford: Oxford University Press, 1994), pp. 239–66.
118. For a gloomy narrative of the nineteenth century, see Mike Davis, *The Late Victorian Holocausts. El Nino Famines and the Making of the New World* (London and New York, 2001).
119. Eric Stokes, 'The Return of the Peasant to South Asian History', *South Asia*, 6(1) (1976): 96–111; for Hardiman's observations on this, see David Hardiman (ed.), *Peasant Resistance in India 1858–1914* (Delhi, 1992), p. 3.

4. The Political Ecology of the Peasant: the Faraizi Movement between Revolution and Passive Resistance

1. Bankim Chandra Chatterjee, 'Bangalir Bahubal', in *Prabandha Pustaka* (Kantalpara, 1880), pp. 1–13. See also Anuradha Roy, *Nationalism as a Poetic Discourse in Nineteenth Century Bengal* (Calcutta, 2003), pp. 135–8.
2. Ranajit Guha, *Dominance without Hegemony: History and Power in Colonial India* (Cambridge, MA, 1997), pp. 205–11.
3. For a sketch of the life and activities of Haji Shariatullah, see Muin-ud-Din Ahmad Khan, *History of the Fara'idi Movement in Bengal, 1818–1906* (Karachi, 1965), pp. 1–22.
4. For an account of the superstitious practices surrounding childbirth, death, marriage and circumcision in contemporary Muslim society and the Faraizi efforts to eliminate them, see James Taylor, *A Sketch of the Topography and Statistics of Dacca* (Calcutta, 1840), p. 240; James Wise, *Notes on the Races,*

Castes, and Trades of Eastern Bengal (London, 1883), pp. 21, 50; Khan, History of the Fara'idi, pp. lxxxvii–lxxxviii.

5. The government identified as many as 23 unauthorized taxes imposed by the zamindars on the peasantry as late as 1872. See Khan, History of the Fara'idi, appendix C. The resentment of the Hindu zamindars against the Faraizi policy of non-payment of puja taxes was reported by the Bengal Police in 1842. See Calcutta Review, 1 (1844): 215–16.

6. Sugata Bose, Peasant Labour and Colonial Capital. Rural Bengal since 1770 (Cambridge, 1993), pp. 49, 148; To some extent, the condition might be compared to that of 'agricultural involution' in colonial Indonesia as described by Clifford Geertz. See his Agricultural Involution: the Processes of Ecological Change in Indonesia (Berkeley and Los Angeles, 1970).

7. Wise, Notes on the Races, pp. 21–2; for other notes on the Faraizi mobilization of the weavers, see Khan, History of the Fara'idi, p.116; C.A. Bayly, The Origins of Nationality in South Asia. Patriotism and Ethical Government in the Making of Modern India (Delhi, 1998), p. 183.

8. Titu Mir was a peasant leader who enlisted significant support in waging resistance against the landlords and the Raj. After initial success, he died in a battle against a British–zamindar coalition in 1831. For a detailed discussion of Titu Mir's life and his influence on subsequent peasant movements, see M.A. Khan, Titu Mir and his Followers in British Indian Records (Dacca, 1977); see also, M.A. Khan, 'Faraizi movement', in Sirajul Islam (ed.), Banglapedia (Dhaka, 2003).

9. Gastrell reported that on Shariatullah's death his followers assembled, and by 'common consent' named Dudu Miyan as the head of the Faraizis. See G.E. Gastrell, Geographical and Statistical Report of the Districts of Jessore, Fureedpore and Backergunge (Calcutta, 1868), p. 36.

10. Muhibuddin Ahmad, Aposhhin ek sangrami pir Dudu Mia [An uncompromising saint leader Dudu Mia] (Dhaka, 1992), pp. 14–15.

11. According to Khan, following the Quranic verse 'whatever is in the heavens and in the earth belongs to God' (Quran, 4:131), Dudu Miyan declared that the land was the bounty of God, and man being His most favoured creature had equal right to exploit this 'divine gift'. Land, therefore, according to the Faraizis, belonged to those who had exploited it. See Khan, History of the Fara'idi, p. 114. This interpretation not only helped the peasantry to perceive the zamindari exploitation in terms of illegal taxes as gross injustice, it legitimized their settlement in the reclaimed and newly formed lands in the Delta; see also Wise, Notes on the Races, p. 24.

12. University Library, Cambridge, Add. MS. 7490/39/4: 'Correspondences relating to the Wahabi movement'.

13. MoE, answer no. 3979.

14. Ibid.

15. Wise, Notes on the Races, p. 25.

16. Gastrell, Geographical and Statistical Report, p. 36.

17. Wise, Notes on the Races, p. 50; J.H. Kramers et al. (eds), The Encyclopaedia of Islam (Leiden: Brill, 1974), p. 784; P.M Holt et al. (eds), Cambridge History of Islam, vol. 2 (Cambridge: Cambridge University Press, 1970), pp. 76–7; Ira M. Lapidus, A History of Islamic Societies (Cambridge, 1988), p. 722; Khan, 'Fara'idi movement', pp. 106–7.

18. IOR, V/27/144/8: The Petition of Bharut Chunder Roy, Mohurir of the Foujdaree Court of Fureedpore, *Trial of Doodoo Meea and His followers 1847*. *Translation of proceedings held in two cases in 1847 before the session Judge of Dacca in which Doodoo Meea and his followers were tried* [henceforth *Trial of Doodo Meea*] (Calcutta, 1848), p. 3.

19. Navinchandra Sen, *Amar Jivan*, in Shajinikanto Das (ed.), *Navinchandra Rachanavali*, vol. 2 (Calcutta, 1366 BS [1959 AD]), pp. 102–10; see also NAB, Dacca Divisional Report, 1880–81, file 120/1/2 B.G.P. Aug 1881; Khan, The 'Fara'idi Movement'.

20. *The Times*, 29 August 1873, p. 7.

21. Isaac Allen, 'The Revival of Islam', *The Calcutta Review* (January, 1874): 47. In most cases, the Faraizis identified themselves as hanafi, sunni or sheikh to avoid any suspicion on the part of the government and therefore it became more difficult to identify them as a distinct community amongst the Muslims. Towards the end of the nineteenth century, government reports often described the Muslims of eastern Bengal as interchangeable with the Faraizis. Hunter reported in the late 1870s, for instance, that the Muslims of the district of Noakhali were nearly all Faraizis. In Comilla their number was 'considerable'. In 1896 one government official reported that the Muslims of Chittagong were 'all stern ferazi'. See W.W. Hunter, *A Statistical Account of Bengal*, vol. IV (London, 1875), pp. 277, 383; NAB, Land Rev 1896, file: Aug 6/a/3 of 1896–7, nos. 56–8, bundle 41, list 17: W.B. Oldham, Comm of Chittagong to Secy to BoR, Lower Provinces, 6 Mar 1896.

22. About the Faraizi quest for autonomy and 'independent statehood', see Nurul H. Choudhury, *Peasant Radicalism in Nineteenth Century Bengal: the Faraizi, Indigo and Pabna Movements* (Dhaka, 2001), pp. 56–61; Bose, *Peasant Labour*, p. 151.

23. 'Faraizi Movement' in *Banglapedia*. The Faraizi violated the legal institutions of the Raj in their everyday life. For instance, they thought that it was justified to give false witness before an English judge. While they considered that lying before a Muslim judge would draw punishment from God, they felt no such fear in lying before a non-Muslim judge. See N.H. Choudhury, *Peasant Radicalism in Nineteenth Century Bengal: the Faraizi, Indigo and Pabna movements* (Dhaka: Asiatic Society of Bangladesh, 2001), p. 59; James C. Scott, *Weapons of the Weak: Everyday Forms of Peasant Resistance* (New Haven, 1985), remains a classic work on the concept of agrarian 'everyday resistance'.

24. Tauriq Ahmad Nizami, *Muslim Political Thought and Activity in India during the First Half of the 19th Century* (Aligarh, 1969), p. 82; see also J.C. Jack, *Survey and Settlement of Bakarganj*, pp. 242–45; Henry Beveridge, *The District of Bakarganj: its History and Statistics* (London, 1876), pp. 255–6; *The Times*, 7 October 1873, p. 9.

25. J. Wise, 'The Muhammadans of Eastern Bengal', *Journal of the Asiatic Society of Bengal*, 1(3) (1894): 33; Nizami, *Muslim Political Thought*, pp. 83–4.

26. University Library, Cambridge, Add. MS. 7490/39/37: 'Memorandum by F.D. Chauntrell, Solicitor to Government of India, on Government's defence against action for wrongful arrest, summarizing recent history of Wahabi Movement in India and previous trials of its members', 12 May 1871, p. 3.

27. A young Dudu Miyan was reported to have sought the blessing of Titu Mir on a visit that he made to him on his way to Mecca. Titu Mir gave him his

tasbih, which Dudu Miyan's descendants still preserve; the connection is also found in the predominance of the Faraizis in the nineteenth century in Barasat, a stronghold of Titu Mir. Possibly they quietly joined the Faraizi rank and file following the end of Titu Mir's resistance.

28. The Faraizi technique of mass mobilization is reflected in the using of Bengali puthi, rather than Persian, Arabic or Urdu language materials. See Kenneth W. Jones, *Socio-Religious Reform Movements in British India* (Cambridge, 1989), pp. 20, 24.

29. *Trial of Doodoo Meea*, p. 65. There are not many sources to determine the extent of Christian involvement in the Faraizi movement. It is possible that some Christians participated individually in the movement. For a general account of the Faraizi influence on non-Muslim communities, see Sen, *Amar Jivani*, pp. 102–3.

30. MoE, answer no. 3979.

31. Although Hunter described the Faraizis as 'intolerant and bigoted', he nevertheless reported that the Faraizis 'never interfere with Hindu religious processions, nor do they annoy the Christian community', Hunter, *Statistical Account*, vol. VI, p. 278; Taylor referred to 'perfect harmony' among the Hindus and Muslims, see Taylor, *A Sketch of the Topography*, p. 257.

32. For notes on the antagonistic relations between the Faraizis and the Muslim landlords, see Choudhury, *Peasant Radicalism*, pp. 70–1.

33. Choudhury, *Peasant Radicalism*, pp. 67–8. For a discussion of the idea of the rich peasant as a mediating agency, see Rajat and Ratnalekha Ray, 'Zamindars and Jotedars: a Study of Rural Politics in Bengal', *Modern Asian Studies*, 9(1) (1975): 97–8.

34. *RAB*, 1871–2, p. 23.

35. APAC, Temple Collections, MSS Eur. F86/165 misc. colln. 14, no. 7: Secy to GoB to the Secy of GoI, 3 Oct 1873.

36. APAC, Temple Collection, MSS Eur. F86/165, misc colln, 14, no.10.

37. For a Marxist approach to the Faraizi movement, see Narahari Kaviraj, *Wahabi and Farazi Rebels of Bengal* (New Delhi, 1982).

38. For the 'riverine character' of the Faraizi movement, and the patterns of the Faraizi settlement along the chars and diaras, see Khan, *A History of the Fara'idi*, pp. 116–22; see also Gastrell, *Geographical and Statistical Report*, p. 36.

39. 'The Exhaustion of the Indian Services', *The Times*, 5 January 1856: 5 (reprinted from *Friends of India*, 8 November 1855).

40. MoE, answer nos. 95–6, 1316; See also RIC, p. IX (343).

41. MoE, answer no. 139. For a description of competition for chars amongst raiyats and planters, see MoE answer nos. 14, 97, 741. When the Indigo Commission asked an indigo planter whether he had ever found the raiyats disinclined to settle on the chars, he replied in the negative. In addition, when settled, he had found the raiyats 'disposed to dispossess him [the planter] always of the best lands', see MoE answer no. 14.

42. Minutes of Richard Temple, RIC, p. XLVII, (381); MoE, answer no. 139.

43. Comm of the Chittagong Div, to Under Secy to GoB, no. 33, 13 Jan 1855, RIC, p. 22.

44. Petition from the inhabitants of Zillah Nuddea to the Lieutenant-Governor of Bengal, 16 February 1860, in RIC, p.186. For peasant perception of unseasonability, see also Offg Magistrate of Nuddea to the Under Secy to the GoB,

no. 31, 12 Jan 1855, in RIC, p. 45; see also MoE answer nos. 482–3, 1121, 1212, 1255,1382, 1392, 2360–61, 2513.

45. Gastrell, *Geographical and Statistical Report*, p. 14; See also Minutes of Richard Temple, RIC, p. XIVII (381).

46. Gastrell, *Geographical and Statistical Report*, p. 14.

47. An example of the humane relationship between a peasant and his cattle is captured in a poignant Bangla short story. See Sharat Chandra Chatterjee, 'Mahesh', in Mushirul Hasan and M. Asaduddin (ed.), *Image and Representation: Stories of Muslim Lives in India* (Delhi, 2000).

48. MoE, answer no. 426.

49. One planter testified to the Indigo Commission that he had prosecuted in a civil court two goalas who used to sow their lands willingly in October, and afterwards destroyed them with their own cattle. See MoE, answer no. 740; this was also because 'raiyats, particularly *goalas* put their cattle into the indigo fields on purpose for the cattle to graze on the doob-grass that grows up with the indigo, which is the very best grass ... they not only do this on a purpose but turn out in large armed bodies, particularly at night, for the purpose of grazing their cattle ... it is no special malice against the planter, but done with a view to get the best grass at a time when the country was burnt up.' See MoE, answer nos. 436, 3567, 2362.

50. MoE, answer nos. 3153, 3154.

51. Nobo Biswas, tahsildar of an indigo concern, testified against some prisoners in *Nobo Biswas vs. Alum Biswas*, Gopal Ghose et al., RIC, p. 267. For Hindu–Muslim cooperation in the indigo movement, see also Dina Bandhu Mitra, *Nila darpana nataka* (Calcutta, 1861).

52. MoE, answer nos. 2962, 3072; see also Dampier's report on the leagues or combinations as formed by the Faraizis: 'Report on the state of the Police in Lower Provinces, for the first six months of 1842' in *The Calcutta Review*, 1 (May–Aug, 1844): 216.

53. Ashley Eden, MoE answer no. 3629.

54. Ashley Eden, Offg Magistrate of Barasat to Comm of Nuddea Div, 6 Apr 1858, RIC, p. 97.

55. R.C. Dutt, *The Peasantry of Bengal*, ed. Narahari Kaviraj (Calcutta, 1980). First published 1874.

56. Blair B. Kling, *The Blue Mutiny: Indigo Disturbances in Bengal, 1859–1862* (Philadelphia, 1966), p. 61; See also David Hardiman, 'Introduction', in David Hardiman (ed.), *Peasant Resistance in India* (Delhi and Oxford, 1992), pp. 14–15.

57. For events leading upto this, see Khan, *A History of the Fara'idi Movement*, pp. 33–4; see also Beveridge, *The District of Bakarganj*, p. 399.

58. James Wise, 'The Muhammadans of Eastern Bengal', *Journal of the Asiatic Society of Bengal*, 1(3) (1894): 51.

59. Prisoner no. 308; see also testimony by Umeerooddeen, prisoner no. 225, in the *Trial of Doodoo Meea*, pp. 182, 193–4.

60. Kling, *The Blue Mutiny*, p. 32.

61. RAB, 1872–3, p. 21.

62. Hardiman, *Peasant Resistance in India*, pp. 17–18.

63. MoE, answer nos. 2390, 2388–9.

64. CSAS, Mukherjee papers, 'Thirty-nine articles', p. 61.

65. Wise, *Notes on the Races*, p. 37.
66. IOR, P/238, Collection 14, no.26/27: A. Abercrombie, Comm of Dacca Div to Secy to GoB, Rev dept, 26 Nov 1873; IOR, P/238, Collection 14, no. 26/27: J.G. Charles, Offg Joint Magistrate of Munshighanj to Collector of Dacca, 19 Sept 1873. For a recent account of the background of conflicts between the reclaiming peasant and the landlord, see Sirajul Islam, 'Permanent Settlement and Peasant Economy', in Sirajul Islam (ed.), *History of Bangladesh* (Dhaka, 1997), pp. 278–80.
67. K.K. Sengupta, 'Agrarian Disturbances in the 19th Century Bengal', in A.R. Desai (ed.), *Peasant Struggles in India* (Bombay, 1979), pp. 189–90.
68. A secretary to GoB compared the eastern Bengal tenants' collective effort to defend their interest with that of the trade unionists in England. They paid heavily to their unions to meet the legal expenditure. See, IOR,Temple Collection, MSS Eur. F86/165, Misc. colln. 14, no. 7: Offg Secy to GoB to Secy of the GoI, 3 Oct 1873. See also Sengupta, 'Agrarian Disturbances in 19th Century Bengal', pp. 195–201; K.K. Sengupta, 'Peasant Struggle in Pabna, 1873: its Legalistic Character', in A. Desai, *Peasant Struggles in India*, pp. 179–88.
69. WBSA, Agriculture dept, no. 1P, 27 Aug 1872, in Sept 1872-169: Offg Secy to GoB, Rev dept to Secy to GoI.
70. RAB, 1871–2, pp. 22–3.
71. These deeds were declared null and void by the civil court, as it found them to be obtained by undue influences. WBSA, collection no.14–10 in Rev dept (L.P), Revenue P.V. Nov 1876–222: P. Nolan, Offg Joint Magistrate, 29 June 1875.
72. Hunter, *Statistical Account*, vol. V, p. 59; also note the quip about Pabna: *ze ashe Pabna tar nei bhabna* ['whoever comes to Pabna nevermore suffers want'], See, L.S.S. O'Malley, *Bengal District Gazetteer: Pabna* (Calcutta,1923), p. 31.
73. Dutt, *The Peasantry of Bengal*, p. xx.; for more recent notes on the Faraizi role in the Pabna movement, see Ranajit Guha, *Elementary Aspects of Peasant Insurgency in Colonial India* (Durham, 1999), p. 173; Hardiman, *Peasant Resistance in India*, p. 21.
74. *The Times*, London, 4 October 1873, p. 10; also note this observation of a government official: 'the case of these petty talookdars is a hard one. They, or at least most of them, declare that they have not the means to go to the Court, and that it is only by borrowing that they will be able to meet the Government demands, and this I believe from their extreme anxiety to come to terms with the ryots', IOR, Temple Collection, MSS Eur. F86/165, misc. colln. 14, no.10–11.
75. WBSA, Collection no. 14-10 in Rev dept (L.P), Revenue P.V. Nov 1876–222P: Nolan, Offg Joint Magistrate, 29 June 1875.
76. RAB, 1874–5, p. 15.
77. Sengupta, 'Agrarian Disturbances', p. 198. For a description of the influence of Noa Miyan on the peasantry and his power vis-à-vis the zamindars, see Sen, *Amar Jivan*, pp. 102–3, 108–10; see also NAB, Divisional Report Dacca, 1880–81, file 120/1/2 B.G.P. Aug, 1881.
78. In his witness to the Indigo Commission, Ashley Eden reported that he heard no single complaint from the tenants of the Sundarbans. MOE, answer no. 3664.

79. MoE, answer no. 2432.
80. MoE, Morrell's answer nos. 2439, 2442.
81. NAB, Rev dept (Land), 'A' progs, wooden bundle 2, list 17: W.E. Morrell, to A. Mackenzie, Offg Secy to GoB, 10 Aug 1869.
82. NAB, Rev dept (Land), 'A' progs, wooden bundle 2, list 17: Petition of the tenants of Tushkhali to the BoR, (1276 BS, 1869 AD).
83. NAB, Rev dept (Land), 'A' progs, wooden bundle no.1, list 17: Robert Morrell, Morrellgunge to Secy to BoR, Lower Provinces 12 Aug 1861.
84. NAB, Rev dept (Land), 'A' progs, wooden bundle no.1, list 17: Robert Morrell, Morrellgunge to Secy to BoR, Lower Provinces 12 August 1861.
85. NAB, Rev dept (Land), 'A' progs, wooden bundle 1, list 17: Reily's report of 8 July 1859, no. 20 quoted in H.L. Dampier, Secy to BoR, to Secy to GoB, 19 Apr 1861.
86. NAB, Rev dept (Land), 'A' progs, wooden bundle no.1, list 17: J.D. Gordon, Junior Secy to GoB, to Offg Secy to BoR, 31 Aug 1861; see also W.E. Morrell to A. Mackenzie, Offg Secy to GoB, 10 Aug 1869.
87. NAB, Rev dept (Land), 'A' progs, wooden bundle 2, list 17: J.C. Price, Offg Collector of Backergunge, to Comm of Dacca Div, no. 1272, 28 Aug 1869.
88. NAB, Rev dept (Land), 'A' progs, wooden bundle 5, list 17: T.B. Lane, Offg Secy to BoR, Lower Provinces to Offg Secy to GoB, Rev dept, no. 7496B, 23 Nov 1869; See also Syed Besarut Ally, Mooktear to Khajah Abdool Gunny, to Secy to GoB, no. 31, 1 May 1869.
89. NAB, Progs for Apr 1867, in Land Rev, 'A' progs, wooden bundle 1, list 17: Rivers Thompson, Offg Secy to BoR, to Secy to GoB, no. 653, 13 Sept 1861.
90. 'It is at the same time quite possible that Mr Reily may have, without any intention, unguardedly allowed expressions to escape his lips, which the turbulent and, for the time being, recalcitrant Furazee [Faraizi] ryots of Tooshkhallee construed to their own advantages', NAB, J.C. Price, Offg Collector of Backergunge, to the Comm of Dacca Div, no. 1272, 28 Aug 1869.
91. NAB, Rev dept (Land), 'A' progs, wooden bundle 2, list 17: F.B. Simson, Comm of Dacca Div, to Offg Secy to BoR, Lower Provinces, no. 22T, 9 Sept 1869.
92. NAB, Land Rev, 'A' Progs, wooden bundle 8, list 17: H. Bell, Offg Legal remembrancer, to Offg Secy to BoR, Lower Provinces. no. 943, 4 Sept 1871.
93. NAB, Land Rev, 'A' Progs, wooden bundle 8, list 17: H. Bell, Offg Legal remembrancer, to Offg secy to BoR, Lower Provinces. no. 943, 4 Sept 1871.
94. NAB, Rev (Land), wooden bundle 15, list 17: H. Beveridge, Collector of Backergunge to Comm of Dacca Div, 8 Dec 1874.
95. NAB, Rev (Land), wooden bundle 15, list 17: Messrs. Morrell and Lightfoot to Collector of Backergung, 10 Sept 1874; WBSA, Revenue P.V. Jan 1875–197: Messrs Morrell and Lightfoot to Secy to GoB, Rev dept; also Messrs Morrell and Lightfoot to Secy to GoB, Rev dept, 16 Dec 1874; WBSA, Rev progs of GoB, P.V., Feb 1876, 210, Messrs Morrell and Lightfoot to Officiating Secy to BoR, Lower Province, 17 Apr 1875: 'We have not been able to collect any portion of the Rs. 15,000 excess revenue put on the farm from the ryots.'
96. NAB, Rev (Land), wooden bundle 15, list 17: Minute by the Lieutenant-Governor of Bengal, 1 Jan 1875.

97. The remnant of the Faraizi resistance in the countryside was curiously non-agrarian and parochial. For instance, the Faraizis were reported to be exceptionally hostile to the programme of vaccination against smallpox. See *East Bengal and Assam Era*, 20 January 1906: 5.

98. For a discussion of the development and flourishing of the Faraizi communication and organizational network along various rivers, see Khan, *A History of the Fara'idi*, pp. 118–20.

99. Some diara surveys were carried out in the 1870s. But the cadastral survey that had started by the turn of the twentieth century truly 'opened up' the countryside. For a detailed discussion of the early colonial drive for exploring the physical landscape of India, see Matthew H. Edney, *Mapping an Empire: the Geographical Construction of British India, 1765–1843* (Chicago, 1997). For an analysis of the relationship between the state power and scientific knowledge of the landscape, see James C. Scott, *Seeing Like a State: How Certain Schemes to Improve the Human Condition have Failed* (New Haven and London, 1998).

100. Dutt, *The Peasantry of Bengal*, p. xxiii.

101. For instance, by the turn of the century, the Khondker, a relatively superior Muslim group, had become averse to carrying any luggage, bags and the like on their heads or shoulders, cultivating their land by themselves and raising clothes beyond their knees. For notes on the 'ashraf's aversion to physical labour', see Rafiuddin Ahmed, *The Bengal Muslims 1871–1906: a Quest for Identity* (Delhi, 1981), p. 11.

102. A.F.M. Abdul Hye, *Adarsha Krishak* [Model Peasant] (Mymensingh, 2nd edn, 1328 BS (1920 AD), preface.

103. No wonder the Nawab of Dhaka was considered 'the most tyrannical landlord'. See oral evidence of the representatives of the Bakarganj Krishak Proja Party, in *Report of the Land Revenue Commission*, Bengal, vol. VI, 1941, p. 364.

104. Sheikh Abdus Sobhan, *Hindu-Mosolman* (Dacca, 1889); Muhammad Dad Ali, *Samaj Siksha* [Instructions on Society] (Nadia, 1910); *Islam-Chitra O Samaj-Chitro* [Sketches of Islam and the Muslim Society dealing with the present degenerate condition of the Muhammadans] (Mymensingh, 1914); Sufia Ahmed, *Muslim Community in Bengal, 1884–1912* (Dacca: S. Ahmed; Distributed by Oxford University Press, Bangladesh, 1974); A.R. Mallick, *British Policy and the Muslims in Bengal 1757–1856* (Dacca, 1977).

105. Sobhan, *Hindu-Mosolman*, pp. 161–4; for a similar argument in favour of a new generation of modern Muslims as against rural mollahs, see Ali, *Samaj Siksha*, p. 95; *Islam-Chitra O Samaj-Chitro*, p. 30.

5. Return of the Bhadralok: the Agrarian Environment and the Nation

1. Partha Chatterjee, 'Agrarian Relations and Communalism in Bengal, 1926–35', in Ranajit Guha (ed.), *Subaltern Studies 1: Writing on South Asian History and Society* (Delhi, 1982), p. 18; Partha Chatterjee's idea of peasant consciousness has found eloquent expression in Taj ul-Islam Hashmi, *Pakistan as a Peasant Utopia* (Boulder, 1992).

2. Sugata Bose, 'The Roots of "Communal" Violence in Rural Bengal: a Study of the Kishoreganj Riots, 1930', *Modern Asian Studies*, 16(3) (1982): 491; see also Sugata Bose, *Peasant Labour and Colonial Capital. Rural Bengal since 1770* (Cambridge, 1993), p. 168.
3. For details of this argument, see Anuradha Roy, *Nationalism as a Poetic Discourse in Nineteenth Century Bengal* (Calcutta, 2003), pp. 123–40.
4. For a first-hand account of an earlier process of the emergence of the 'middle class' in the Calcutta urban space, see Bhabani Charan Bandyopadhyaya, *Kalikata Kamalalaya* (Calcutta, 1938).
5. Amalendu De, 'Sri Aurobido's Role in Indian Freedom Struggle: an Assessment from Different Perceptions', Presidential Address 2002–03 (Kolkata, 2003).
6. Prathama Banerjee, 'The Work of Imagination: Temporality and Nationhood in Colonial Bengal', in Shail Mayaram, M.S.S. Pandian and Ajay Skaria (eds), *Subaltern Studies XII: Muslims, Dalits and the Fabrications of History* (New Delhi, 2005), pp. 295–6.
7. Pitambar Sarkar, *Jati-Vikash* [Evolution of Caste] (A Treatise on the Caste System in Bengal, with Reference to the Caste Name Chasadhola) (Calcutta, 1910), pp. 136–7.
8. Sarkar, *Jati-Vikash*, p. 136.
9. Jadunath Majumdar, *Hindu Samajer Samashya* (Jessore, 1325 BS), pp. 56, 66–70.
10. Pandit Rajanikanta, *Swadeshi Palli Sangeet* [Swadeshi Village Songs] (Mymensingh: 1907).
11. Ramhari Bhatcharya, *Banchibar Upay* [*Essays on the Bread Problem and other Economic Issues*] (Jessore, 1332 BS, 1925 AD), pp. 20–5.
12. For a note on different agro-ecological scenarios in West and East Bengal, see Sugata Bose, 'Agricultural Growth and Agrarian Structure in Bengal: a Historical Overview', in Ben Rogaly, Barbara Harriss-White and Sugata Bose (eds), *Sonar Bangla? Agricultural Growth and Agrarian Change in West Bengal and Bangladesh* (New Delhi, 1999), pp. 44–5.
13. NAB, GoB, Agriculture and Industries dept (Agriculture), March 1927, Progs no. 1–20, Royal Commission on Agriculture in India in NAB, Archives file no. 55: Memorandum by the Director of Agriculture, Bengal, regarding rainfall of each district, its relation to the system of cropping in vogue.
14. Brindabon Chandra Putatunda, *Nuton Banger Puraton Kahini* [Old Stories of New Bengal] (Bakarganj: 1323 BS), p. 103.
15. Rishikesh Sen, *Bekar Samasya* [An Essay on the Causes and Cure of Unemployment] (Chandannagor, 1934), pp. 83–5, 152–4.
16. *Report on the Census of India*, 1901, vol. 6: Bengal, part 1 (Calcutta, 1902), p. 351.
17. *Report of the Census of India*, 1931, vol. 5: Bengal, part 1 (Calcutta, 1933), p. 454.
18. M.K.U. Molla, *The New Province of Eastern Bengal & Assam, 1905–1911* (Rajshahi, 1981).
19. *Collected Works of Mahatma Gandhi*, vol. 5 (Delhi, 1984), p. 121.
20. CSAS, Pinnell Papers: 'With the sanction of government', Memoir of L.G. Pinnell (published privately by M.C. Pinnell in 2002), p. 63.
21. *Bengal Legislative Council Proceedings* [hereafter BLCP], 35, 3 (1930), 468.

22. NAB, Progs of Government of Bengal, Revenue dept (Land), nos. 19–20, Mar 1922, p. 16.
23. NAB, Government of Bengal, Rev dept (Land), no. 6, Apr 1922, p. 6. Coincidentally, in a famous Bangla novel, there is a character, Balaram Kaviraj, who was settled in a coastal island. He was depicted in the novel in the following words: 'In this ... *char*, he has created his own world. There is not much anxiety for patients. There are plenty of lands in the char. There are ponds of saline water as well as gardens of betel nuts. He also owns about fifty buffalos – he can be regarded as a little zamindar. Therefore, *Kaviraji* (medical practice) may be regarded as his *nesha* (obsession) rather than *pesha* (profession).' See Narayan Gangopadhaya, *Narayan Gangopadhaya Rachanabali: Upanibesh* [*The Colony*], vol. 2 (2nd edn, Calcutta, 1386 BS, AD 1980), p. 227.
24. *Report of the Bengal Provincial Banking Enquiry Committee* [hereafter BPBEC], *1929–30*, vol. 2, Evidence, part 1(Calcutta, 1930), p. 248.
25. BLCP, 41, 3 (1933), 63–4.
26. *Annual Report on the Working of Co-operative Societies in the Presidency of Bengal* [hereafter, RWCSB] *(1912–13)*, 20; see also RWCSB (1924–5), 21.
27. NAB, 'Settlement of Khas Mahal land in Faridpur with young men of the bhadralok class', progs of Rev dept (Land), no. 1–2, Jan 1928, in Jan–Mar 1928.
28. Rajshahi Commissioner's Library, Rajshahi, 'Settlement of Khasmahal Land with "Bhadraloke" Youths in Faridpur', 1933.
29. *Bengal Waste Lands Manual 1919* (Calcutta, 1919), p. 2.
30. *Bengal Government Estates Manual, 1932* (Calcutta, 1933), p.19; see also the case of *Rajani Kantha v. Yusuf Ali*, referred to in Rai Surendra Chandra Sen Bahadur, *The Bengal Tenancy Act* (5th edn, Calcutta, 1925), p. 76.
31. *Bengal Waste Lands Manual 1936* (Calcutta, 1936), p. 1.
32. *Bengal Legislative Assembly Proceedings* [henceforth *BLAP*], 54, 2 (Calcutta, 1940), p. 131.
33. Pomode Ranjan Das Gupta, *Faridpur Revisional Settlement Final Report, part-1 – Survey and Settlement Operations 1940–4* (Dacca, 1954), p. 5.
34. Krishnakali Mukeerji, 'The Transferability of Occupancy Holding in Bengal', part I, *The Bengal Economic Journal*, 1 (3) (1917): 263.
35. Peter Marshall, *Bengal: the British Bridgehead*, The New Cambridge History of India (Cambridge: Cambridge University Press, 1988, reprinted 1990), pp. 22–3.
36. Krishnakali Mukeerji, 'The Transferability of Occupancy Holding', 266.
37. NAB, Rev (Land) 'A' progs (confidential), bundle 1, list 124D, 1939, p. 2: 'Memorial of landholders in the matter of the Bengal Tenancy Act, 1938'.
38. NAB, 'A' progs, wooden bundle 26, list 17, Mis – Collection, XIV, no. N100–101, Aug 1889: C.E. Buckland, offg Secy to BoR, Lower Provinces, to Secy to GoB, no. 5414, 3 Jul 1889.
39. NAB, Progs of the Rev dept (Land) for May 1927 (Bengal Tenancy Act Amendment Bill), pp. 218–19: A. Chaudhuri and B. Chakravarti, Honorary Secretaries, Bengal Landholder's Association, to Secy, GoB, 2 July 1923.
40. NAB, 'A' Progs, wooden bundle no. 26, list 17, Mis – Collection, XIV, no. N100–101, Aug 1889: C.E. Buckland, Offg Secy to BoR, Lower Provinces, to Secy to GoB, R.D., no. 5414, 3 July 1889.

41. NAB, Rev dept (Land), 'A' Progs, bundle 76, list 17, June 1921, file L.R. 2-A – 4(1) of 1925, in progs nos. 7–8, p. 64; See also NAB, Government of East Bengal (Land Rev), confidential, 'A' progs, bundle I, list 124, file no. L.R. 1-A – 54, serial 1, Rev dept (Land) 1938.
42. Bose, *Peasant Labour*, p. 124.
43. NAB, Government of East Bengal (Land Rev), confidential, 'A' progs, bundle 1, list 124B: note by Sir F. Sachse on the 'Report of the rent law commission, 1880, and its draft bill section 20 recommended transferability for raiyats', p. 5 in confidential file no. L.R. 1-A – 54, serial 1, GoB, Rev dept (Land) 1938.
44. Bengal Tenancy Act of 1939 'facilitated the transfer of under-*raiyati* holdings and enabled these rights also to be purchased by non-agriculturists', See Adrienne Cooper, *Sharecropping and Sharecroppers' Struggles in Bengal 1930–1950* (Calcutta, 1988), p. 47.
45. In the nineteenth century legal disputes revolved around the accrual of occupancy rights; in the twentieth century these shifted to issues of 'transfer of occupancy holdings by involuntary sales'. See NAB, Babu M.M. Deb, Secy, Tippera People's Association, to Secy to GoB, Rev dept, 8 Jan 1915, in Progs of GoB, Rev dept (Land), June 1916.
46. NAB, Progs of GoB, Rev dept (Land), FQE, June 1927, pp. 359–60: the second part (paragraphs 11 to 20) of the opinions of the Cultivators Association, Lakhipur and Raipur (Noakhali) on the Preliminary draft of the Bengal Tenancy (Amendment) Bill, no. 92–93, file: 2-A-1(67) of 1923.
47. NAB, Rev dept (Land), progs 'A', June 1916, p. 115: Maulvi Wasimuddin Ahmed, Secy, Anjuman-I-Islamia to Chief Secy to GoB, 9 Jan 1914.
48. NAB, Progs of Rev dept (Land), April-June 1916, p. 89: F.C. French, Offg Comm of Dacca Div, to Secy to GoB, Rev dept, 4 Jan 1915.
49. NAB, Progs of Rev dept (Land), April–June 1916, p. 92: F.W. Strong, Collector of Bakarganj to Comm of Dacca Div, 3 Dec 1914.
50. B.C. Prance, *Final Report on the Survey and Settlement Operations in the Riparian Area of District Tippera Conducted with the Faridpur District Settlement 1909 to 1915* (Calcutta, 1916), p. 25.
51. NAB, Judicial dept, list 114, bundle 2, no. 1-8-2642. file 55C – 13/33(1), pp. 1–3: Rai R.C. Sen Bahadur, Land Rev Settlement Officer, Chittagong, to Director of Land Records, Bengal, 2 Aug 1933.
52. On the suggestion that 'time to return has arrived', see Sri Baidyanath's introductory remarks to Ramhari Bhatcharya, *Banchibar Upay.*
53. Baden-Powell noted in the late nineteenth century that agriculture was 'positively abhorred' by the Brahmans and Kshatriyas, who comprised the bulk of the bhadralok, although the third castes or Vaishya (common people) became 'to some extent an agricultural class'. See, B.H. Baden-Powell, *The Origin and Growth of Village Communities in India* (London, 1899), p. 52.
54. Biseshwar Bhatcharya, 'Gramer Kotha' [Thoughts on the village], *Bangabani*, 4 (Baishakh 1332 BS, AD 1926), pp. 270–1.
55. Jogesh Chandra Roy Bidyanidhi, '*Annachinta* [Thoughts on Subsistence]', *Prabashi*, vol. 25, part. 1, no. 6. (Ashwin, 1332 BS, AD 1926).
56. Report of the Land Revenue Commission (Floud Commission Report), vol. 1, 1940, p. 37. (Henceforth, LRC.)

57. Santipriya Basu, *Banglar Chashi* (Calcutta, 1351 BS, AD 1994), pp. 38–40.
58. NAB, Progs of Rev dept (Land) for May 1927 (Bengal Tenancy Act Amendment Bill, 1923): S.C. Chakraverty, Attorney-At-Law, Joint Secy, Bengal Mahajan Sabha to Secy, GoB, Legislative dept, 23 May 1923.
59. NAB, Progs of GoB, Rev dept (Land), April 1923, p. 54: Babu Kshetra Mohan Roy, Pleader and Zamindar, Comilla to Collector of Comilla.
60. NAB, Progs of the GoB, Rev dept (Land), p. 155: Babu Sarat Chandra Ray, Senior Government Pleader, Rajshahi to District Judge of Rajshahi, 12 Feb 1923.
61. NAB, Progs of GoB, Rev dept (Land), April–June 1927, p. 234: Babu Kaliparasanna Guha Chowdhury, Pleader and Zamindar, Honorary Secy, Bakarganj Land Holders' Association, to Secy to GoB, Rev dept, no. 53, 18 Apr 1923.
62. NAB, Progs of GoB, Rev dept (Land), May 1927, p. 58: 'Opinion on the Bill for the amendment of the Bengal Tenancy Act' by Rai Satish Chandra Sen Bahadur, Senior Government Pleader, Chittagong.
63. NAB, Progs of GoB, Rev dept (Land), May 1927: J.F. Graham, District Judge Dacca to Assistant Secy, GoB, Rev dept, 15 May 1923.
64. NAB, Progs of GoB, Rev dept (Land), April–June 1927, p. 158: Babu Birendar Chandra Sen-Gupta, Munsif of Rampur-Boalia to District Judge, Rajshahi, 10 Apr 1923. On this point, see also Mohiuddin Mondal, President of a meeting of 'Jotedars and raiyats' of Rajshahi to Secy to GoB, Rev dept, no. 77, 16 Apr 1923, NAB, Progs of GoB, Rev dept (Land), April–June 1927, p. 320.
65. Radharomon Mukherjee, *History and Incidents of Occupancy Right* (Delhi, 1919, reprinted 1984), p. 10.
66. Mookerjee, *History and Incidents of Occupancy Right*, pp. 114, 137.
67. NAB, Land Revenue, A proceeding, Bundle 57, list 17: C.J. Stevenson-Moore, Inspector General of Police, Lower Province to the Secretary to the BoR, Lower Provinces, in file no. 22-R/7 – 16 of 1905: Raiyatwari Settlement in the Sundarbans, GoB: Revenue dept (land revenue), no. 117–21, p. 14.
68. NAB, Revenue dept in Progs of the GoB, Revenue dept (Land Revenue), June 1916: H.M. Haywood, Secretary to Bengal Chamber of Commerce to Secretary, GoB, Revenue Department in Proceedings of the GoB, Revenue Department (Land Revenue), June 1916.
69. S. Bose, *Peasant Labour*, pp. 122, 131; Adrienne Cooper, *Sharecropping and Sharecroppers' Struggle*, pp. 91–4.
70. Shantipriya Basu, *Banglar Chashi*, pp. 26–7. For a study of indebtedness in Bengal, see Sirajul Islam, 'The Bengal Peasantry in Debt, 1904–1945', *Dacca University Studies*, vol. XXII (part A) (1974). It may be noted that at a gathering of the peasants of Eastern Bengal and Assam held in 1937, the papers simply listing the names of the indebted peasants weighd about 20 maunds and 27 seers (about 800 kilograms). Quoted in Syed Abul Maksud, *Maolana Abdul Hamid Khan Bhasani* (Dhaka, 1994), p. 28.
71. Binay Bhushan Chaudhuri, 'The Process of Depeasantization in Bengal and Bihar, 1885–1947', *Indian Historical Review*, 21(1) (1975).
72. Sugata Bose, *Agrarian Bengal: Economy, Social Structure and Politics, 1919–1947* (Cambridge, 1986), p. 102.
73. BPBEC, p. 233.

74. BPBEC, p. 333.
75. S. Islam, 'Bengal Peasantry in Debt', p. 59.
76. Henry W. Wolfe, *People's Banks: a Record of Social and Economic Success* (3rd edn, London, 1910), p. 520. The idea that the East Bengal people possessed enormous amount of 'idle' capital does not appear entirely unfounded given the relative prosperity of the region throughout the nineteenth century. In describing the 'effect of the commercial and social changes on the mass of the people' in East Bengal, a *Times* reporter reported in 1873, referring particularly to the Mymensingh district, that in the previous three years the peasantry had received a million pounds sterling for jute alone. Given the fact that the cultivators did not indulge in fancy clothing, ornaments, food or finer houses and that there were only few usurers, the reporter wondered where did that huge amount of money go. He found out that the money was 'carefully buried in the mud floor' of the peasant's house or in the bag which was 'covered up in the corner of family box', *The Times*, 7 October 1873.
77. J.P. Niyogi, *The Co-Operative Movement in Bengal* (London, 1940), p. 62.
78. RWCSB, year ending 30 June 1937, Summary of General Progress, TBCJ, XXV, 1 (1939), p. 57.
79. Jessore Collectorate Library, Jessore, shelf mark XXIXW: R. Gourlay, to L. Hare, Member, BoR, Lower Provinces, Dec 1904, p. 20.
80. RWCSB, 1911–12 (Calcutta, 1912), p. 11.
81. Hireankumar Sanyal, 'Co-operation in Bengal', *BCJ*, XVIII(4) (1933), p. 151.
82. NAB, GoB, dept: Co-operative, Progs: B, Bundle 1, List 37, file no. 8: Bakarganj Co-Operative Central Bank Limited, 16th Annual Report (in Bengali), Bakarganj, 1928–29, p. 14.
83. A. Sadeque, 'The Co-operative Credit Movement and Interest Rate in India', *TBCJ*, XXV(2) (1939): 22; see also Annual Report on the Working of Co-operative Societies in the Presidency of Bengal for the Year Ending 30 June 1937, Summary of General Progress, *TBCJ*, XXV(1) (1939): 54–5.
84. NAB, CCRI (R–I), 'B' Progs, Bundle 1, List 36: Joint Secretary's note, in GoB, 1937, CCRI dept (R–I), file 2M-40, in NAB, CCRI (R–I), 'B' Progs, Bundle: 1, List: 36.
85. NAB, CCRI dept (Rural Indebtedness), 'B' Progs, bundle 9, list 36. file 4C-8/42: Deputy Director, Debt conciliation, Burdwan Div, to Assistant Secy to GoB, CCRI dept, 14 June 1942.
86. D.L.M., 'The Bengal Agricultural Debtors' Act: the Position of Co-operative Societies – A Further Note', *TBCJ*, XXII(2) (1936): 57, 62.
87. 'Some Stray Thoughts on Co-operation in Bengal', *The Bengal, Bihar, Orissa Co-operative Journal*, IX(IV) (April 1924): 311.
88. NAB, GoB 1935, Co-operative dept of, Progs B, Bundle 1, List 37(1), file no. 62.
89. BPBEC, p. 39.
90. BPBEC, p. 486.
91. Abinash Chandara Nag, *Krishak o Sramajibi: Things to Know about Co-operative Credit Societies* (Calcutta, 1907), p.167.
92. S.C. Mitter, *A Recovery Plan for Bengal* (Calcutta, 1934), p. 188.
93. 'Swadeshi Movement', *Eastern Bengal and Assam Era*, 3 Jan 1906, p. 6.
94. P.J. Hartog, in the *Report of the Government of Bengal Unemployment Enquiry Committee*, vol. II, Written and Oral Evidence (Calcutta, 1925), p. 22.

95. NAB, Co-operative dept, 'B' Progs, bundle 1, list 37, file no. 8, pp. 7–9: *Report on the Audit of the Bakarganj Co-operative Central Bank Limited for the Year 1931–32.*
96. Rajkumar Chakravarti and Ananga Mohana Dasa, *Sandwiper Itihasa* (Calcutta, 1330 BS, 1922 AD), pp. 126–7, 166.
97. F.D. Ascoli, *A Revenue History of the Sundarbans from 1870–1920* (Calcutta, 1921), p. 28.
98. M. Rahman, *Sundarbane Ekbasar* [One Year in the Sundarbans], *The Azad*, 18 and 22 December 1936.
99. Niyogi, *Co-operative Movement*, pp. 3–4.
100. NAB, GoB, Agriculture dept (Co-operative), list 37, bundle 1, file 48: D.M. Hamilton, 'Saint Andrew & the Man Standard', paper read by Sir Daniel Hamilton at the University of Calcutta on 10 Jan 1934.
101. B. Roy, ed., *Speeches and Addresses of Sir John Anderson, 1932–1937* (London, 1939), pp. 391–2.
102. LRC, vol. V, p. 371.
103. Ian J. Catanach, *Rural Credit in Western India 1875–1930: Rural Credit and Co-operative Movement in the Bombay Presidency* (Berkeley, 1970), pp. 1–2. Also note this remark: 'There is on the whole a perceptible tendency in the rural societies towards politics. Whether such a tendency is to be checked, it is for others more intimately connected with the movement to judge, but I would like to see the movement develop on economic lines at the present moment.' Presidential address to Rajshahi Division Co-operative Conference, *TBCJ*, XII(3) (1927): 321.

6. The Railways and the Water Regime

1. Karl Marx, 'The Future Results of British Rule in India', reprinted in Ian J. Kerr (ed.), *Themes in Indian History. Railways in Modern India* (Oxford, 2001), pp. 62–7.
2. M.K. Gandhi, 'The Condition of India: Railways', reprinted in Kerr, *Railways in Modern India*, pp. 77–80; see also David Arnold, *Science, Technology and Medicine in Colonial India* (Cambridge, 1987), p. 127.
3. Hena Mukherjee, *The Early History of the East Indian Railway 1845–1879* (Calcutta: Firma KLM, 1994); Mukul Mukherjhee, 'Railways and their Impact on Bengal's Economy: 1870–1920', *Indian Economic and Social History Review*, 17(2) (1980): 191–209; I.D. Derbyshire, 'Economic Change and the Railways in North India, 1860–1914', *Modern Asian Studies* 21(3) (1987): 521–45; Daniel Thorner, 'Capital Movement and Transportation: Great Britain and the Development of India's Railways', *Journal of Economic History*, 11(4) (1951): 89–402.
4. Ravi Ahuja, ' "The Bridge-Builders": Some Notes on Railways, Pilgrimage and the British "Civilizing Mission" in India', in Harald Fischer-Tine and Michael Mann (eds), *Colonialism as Civilizing Mission: Cultural Ideology in British India* (London, 2004), pp. 95–115. For an interesting study of the place of railways within the discourse of modernity and the national imagination and practice, see Laura Bear, *Lines of the Nation: Indian Railway Workers, Bureaucracy, and the Intimate Historical Self* (New York, 2007).

5. Ira Klein, 'Malaria and Mortality in Bengal 1840–1921', *Indian Economic and Social History Review*, 9(2) (1972): 132–60.
6. 'Railways in India. Bengal', *The Colonial Magazine & East India Review*, 16(65) (1849): 349–50; An Old Indian Postmaster, *Indian Railways; As Connected with the Power and Stability of the British Empire in the East etc.* (London, 1846); J.P. Kennedy, *A Railway Caution! Or Exposition of Changes Required in the Law & Practice of the British Empire etc.* (Calcutta, 1849).
7. Transit, *A Letter to the Shareholders of the East Indian Railway and to the Commercial Capitalists of England and India* (London, 1848), p. 8.
8. K.G. Mitchell and L.H. Kirkness, *Report on the Present State of Road and Railways Competition and the Possibilities of their Future Co-ordination and Development, and Cognate Matters, in Governors Provinces* (Calcutta, 1933), p. 6. The Bengal and Assam Railway comprised 3485 miles on 31 March 1945. See M.B.K. Malik, *Hundred Years of Pakistan Railways* (Karachi, 1962), p. 18.
9. *A Sketch of Eastern Bengal with Reference to its Railways and Government Control* (Calcutta, 1861), pp. 13, 21–2, 49.
10. O.C. Lees, *Waterways in Bengal: Their Economic Value and the Methods Employed for Their Improvement* (Calcutta, 1906), p. 9.
11. NAB, Communication, Building and Irrigation (henceforth CBI) dept (Railway), 30 October 1928, bundle 1, unrecorded files, file 7: N. Pearce, Agent, Eastern Bengal Railway to G.G. Day.
12. The Eastern Canals comprised several natural and artificial waterways which connected eastern Bengal to Kolkata port.
13. *A Sketch of Eastern Bengal*, pp. 10–11.
14. For details of the arguments and counter-arguments, see R. Cort, *The Anti-Rail-Road Journal; or, Rail-Road Impositions Detected* (London, 1835).
15. Cort, *The Anti-Rail-Road Journal*, pp. 51–6.
16. APAC, MSS Eur F290/33: *Railway Consolidation Act, 1845*, p. 360.
17. Act V of 1864 (The Bengal Canal Act, clauses 3–4) thus reads: 'It shall be lawful for the Lieutenant Governor of Bengal ... to authorize ... to make and open any navigable channels, or to clear and deepen any navigable channel and to stop any nagivable channel ... no action or suit shall be brought against the State in respect to any injuries or damage caused by or resulting from any act done.' Quoted in B.B. Mitra, *Laws of Land and Water in Bengal and Bihar* (Calcutta, 1934), p. 250.
18. *Rules for the Preparation of Railway Projects with Notes.* In Railway Department, Technical Paper no. 192: G. Richards, Chief Engineer with the Railway Board of India, in enclosure to letter no. 65, 3 February 1919, p. 16.
19. One of the reasons why some officials in the East India Company were initially hesitant over constructing the railways was the apprehension that periodical inundation, among other 'Indian problems', would pose a particular threat to the stability of the same. See 'Indian Railways', *The Times*, 13 August 1846, p. 8.
20. S. Finney, *Railway Construction in Bengal. Three Lectures Delivered at the Sibpur Engineering College in January–February 1896* (Calcutta, 1896), p. 5.
21. H.W. Joyce, *Five Lectures on Indian Railway Construction and One Lecture in Management Control* (Calcutta, 1905), p. 20.
22. Finney, *Railway Construction in Bengal*, p. 8.
23. Finney, *Railway Construction in Bengal*, p. 10.

24. Joyce, *Five Lectures*, p. 55.
25. Joyce, *Five Lectures*, pp. 59–60. For a note on the colonial engineers as 'the handmaidens of capitalist enterprise, charged with turning profit from water', see William Beinart and Lotte Hughes, *Environment and Empire* (Oxford, 2007), p. 147.
26. NAB, CBI dept (Railway), bundle 1, list 70, file 1W-4/1939: P.C. Roy, Superintendent Engineer to Chief Engineer, Communication and Works dept, 14 March 1940.
27. NAB, Progs of the Irrigation dept, April–June 1927: *Report on the Hydraulic Conditions of the Area Affected by the North Bengal Floods*.
28. Finney, *Railway Construction*, p. 8.
29. For instance, the mouth of the Bamanidah river was closed by the embankment of the Eastern Bengal Railway. See NAB, Progs of Irrigation dept, nos. 15–17, January–March 1930, file 13R-8.
30. *The Times*, 13 October 1922, p. 11. For the first major account of the impact of railway embankments on the water system and agriculture in Bengal, see C.A. Bentley, *Malaria and Agriculture in Bengal. How to Reduce Malaria in Bengal by Irrigation* (Calcutta, 1925).
31. Finney, *Railway Construction*, p. 10.
32. Some geologists suggest that Chalan beel is an abandoned bed of the river Ganga (Padma). See, Bisheswar Bhattacharya, 'Bange Ganga' [The Ganga in Bengal], *Bangabani*, 3(2) (1331 BS, AD 1925): 171.
33. M. Abu Hanif Sheikh et al., *Chalan Beel Anchaler Nadimalar Shankyatattik Bissleshon* [Statistical analysis of the rivers of the Chalan Beel], *Institute of Bangladesh Studies Journal*, 7(1) (1406 BS, AD 2000): 125, 129.
34. For a description of the process, see *Report on the Hydraulic Condition of the Area Affected by the North Bengal Floods*, in Progs of Government of Bengal, Irrigation dept, Irrigation Branch, for the quarter ending June 1927, vol. IV.
35. *Report on the Hydraulic Condition of the Area Affected by the North Bengal Floods*, pp. 6–7.
36. W.H. Nelson, *Final Report on the Survey and Settlement Operation in the District of Rajshahi, 1912–1922* (Calcutta, 1923), p. 8.
37. NAB, Progs ('B') of the CBI dept (Railway), bundle 2, list 70, file 1W-1/(19)46: B.M. Mukherjee, Executive Engineer, Khulna Division, to Superintendent Engineer, 21 June 1945.
38. A long list of culverts and bridges with inadequate space for passing of water is provided in NAB, Progs ('B') of the CBI dept (Railway), bundle 2, list 70, file 1W-1/(19) 46.
39. Superintendent of Agriculture, Rajshahi Division, to Deputy Director of Agriculture, Northern Circle, Public Health dept, 24 November 1922, in NAB, 'B' Progs, nos. 156–57, bundle 5, list 4, file P.H.4A-14/1925. For a description of the contribution of railway embankments to the 1918 flood, 'heaviest of all' in north-eastern Bengal, see NAB, Progs of Revenue dept, July 1925: *Final Report on the Flood of Rajshahi Division During the year 1922*, pp. 1–2.
40. NAB, 'A' Progs of Revenue dept, nos. 9–10, wooden bundle 33, list 14, file 6-F-1/ 1919: J.T. Rankin, Commissioner of Rajshahi, to Secretary to Govt of Bengal, Revenue dept, 25 November 1919.

41. NAB, Progs ('B') of Public Health dept, nos. 156–7, list 4, bundle 5, file P.H.4A-14: Superintendent of Agriculture, Rajshahi Division, to Deputy Director of Agriculture, Public Health dept, 24 November 1922.

42. NAB, CBI dept (Railway), unrecorded files, bundle 1, list 70, file 122-5/1941: 'Taking Afflux Observations on Bengal and Assam Railway Water Openings Between Kishorganj and Bhairab-Bazar'.

43. BLAP, LVI, (1940), pp. 92–3.

44. NAB, CBI dept (Railway), 'B' Progs, bundle 2, file 13C-6/194: 'Extracts from notes and orders of Railway Branch', 3 September 1943.

45. NAB, Revenue dept (Land), 'A' Progs, bundle 1, list 90 (A), file 80-R-7H, *Final Report on the Relief Operation in the District of Mymensigh During the Years 1931–1932*.

46. NAB, Progs of Irrigation dept, no. 22, April–June 1926: R.C. Sen, Subdivisional Officer, Brahmanbaria, to Collector, Tippera.

47. NAB, Progs ('B') of Public Health dept, March 1925, nos. 156–157, bundle 5, list 4, file P.H.4A-14: Superintendent of Agriculture, Chittagong, to Deputy Director of Agriculture; BLAP 62, 1 (1942), p. 76.

48. NAB, Proceedings ('A') of Agriculture dept, bundle 29, list 14, file 95: *Final Report on the Relief Operations in the District of Noakhali in 1915*.

49. Kiran Chandra De, *Report on the Fisheries of Eastern Bengal and Assam* (Shillong, 1910), p. 71.

50. C. Addams-Williams, *Notes on the Lectures of Sir William Willcocks on the Irrigation in Bengal Together With a Reply by Sir William Willcocks* (Calcutta, 1931), p. 10.

51. Kiran Chandra De, *Report on the Fisheries*, p. 71.

52. This narrative of the construction of the Hardinge Bridge is heavily drawn from *Hooghly to the Himalayas. Being An Illustrated Handbook to the Chief Places of Interest Reached by the Eastern Bengal State Railway* (Bombay, 1913), pp. 25–6.

53. Addams-Williams, *Notes on the Lectures of Sir William Willcocks*, p. 10.

54. BLAP LIII, (1937), 375.

55. BLAP LV, (1939), 142–3.

56. Seith Drucquer, 'On the Rivers in East Bengal', *Geographical Magazine* 12(5) (March 1941): 352.

57. H.S.M. Ishaque, *Agricultural Statistics by Plot to Plot Enumeration in Bengal 1944 and 1945* (part II, Alipore, 1947), pp. 8–9.

58. For a discussion of the elite and middle-class approval of the railways, see Iftekhar Iqbal, 'The Railway in Colonial India: Between Ideas and Impacts', in Roopa Srinivasan et al. (eds), *Our Indian Railway: Themes in India's Railway History* (Delhi, 2006), pp. 175, 180–3.

59. *Report on the Embankments of the Rivers of Bengal* (Calcutta, 1846), pp. 1–4.

7. Fighting with a Weed: the Water Hyacinth, the State and the Public Square

1. The presence of the plant was first remarked in Florida in the 1890s, in Queensland, Australia, in 1895, in South Africa in 1900, in Cochin China in

1908 and in Myanmar in about 1913. By this time, it had also invaded most of southern and northern Africa.

2. Kabita Ray, *History of Public Health: Colonial Bengal 1921–1947* (Calcutta, 1998), p. 284; see also Seith Drucquer, 'On the Rivers in East Bengal', *Geographical Magazine* 12(5) (March 1941): 3501.

3. Drucquer, 'On the Rivers of East Bengal', p. 350. This theory seems quite plausible since there is evidence that towards the end of the nineteenth century the number of visitors to Calcutta Botanic Garden increased to such an extent that the authorities found it difficult to deal with cases of violation of the garden rules. This made it necessary to ask the government for sanctioning certain additional rules. See *Annual Report of the Royal Botanic Garden* (Calcutta, *1897–8*), 1.

4. NAB, 'Council Question', March 1921, Agriculture and Industries dept (hereafter, AID) (Agriculture Branch), bundle 34, list 14, file 6.

5. About the German 'conspiracy', see Kabita Ray, *History of Public Health*, pp. 284–5.

6. For a note on plant research and plant migration by private companies, see William Neinary and Karen Middleton, 'Plant Transfers in Historical Perspective: a Review Article', *Environment and History*, 10 (2004): 13.

7. NAB, AIB (Agriculture Branch), 1921, 'A' Progs, bundle 34, list 14, file 32.

8. *Amrita Bazar Patrika*, 26 March 1926, 4.

9. J. Chaudhuri, 'Agrarian Problems of Bengal – 1', *The Bengal Cooperative Journal*, XXI (January–March 1936): 117.

10. Radhakamal Mukerjee noted in the early 1930s that a 'few prickly pears introduced into Eastern Australia and water hyacinth into the delta of Eastern Bengal – both as botanic curiosities – have now covered thousands of miles and become a serious menace to agriculture, and communications'. See Radhakamal Mukerjee, 'An Ecological Approach to Sociology', in Ramachandra Guha (ed.), *Social Ecology* (Delhi, 1994), p. 25.

11. BLAP LIII (1938), 224–5.

12. BLAP LIV (1939), 93; also see BLAP LV (1940), 273–4.

13. BLAP LIV (1939), 32–3.

14. 'Lilac Devil in Bengal', *ABP*, 21 August 1928, 12.

15. Benoyendranath Banerjee, 'Some Economic Problems of Bengal – 1: The Water Hyacinth', *The Bengal Cooperative Journal*, XIX (1933): 32, 35.

16. Chaudhuri, 'Agrarian Problems of Bengal – 1', 117.

17. Mostopha Kamal Pasha, 'Water Hyacinth', *Banglapedia: National Encyclopaedia of Bangladesh* (Dhaka, 2002).

18. 'A Note on the Water Hyacinth', by Kenneth McLean, Secretary to Water Hyacinth Committee, in *Report of the Water Hyacinth Committee* (Calcutta, 1922) (hereafter, *RWHC*), XVI. See also M.C. McAlpin's note of 16 April 1919, in NAB, Revenue dept (Agriculture), 'A' Progs, nos. 24–27, July 1919.

19. BLAP LIV (1939), 458–60; BLAP LIII (1938), 378–80.

20. 'Geographical Records', *Geographical Review*, 36(2) (1946): 329.

21. Babu Nibaran Chandra Das Gupta, BLCP I, 3 (1921), 76–7.

22. Recent researches have proved that *V. cholerae* are found to concentrate on the surface of the water packed with the water hyacinth. See W.M. Spira et al., 'Uptake of *V. cholerae* biotype El Tor from Contaminated Water by Water

Hyacinth (*Eichornia Crassipes*)', *Applied Environmental Microbiology*, 42 (1981): 550–3.

23. NAB, AID (Agriculture Branch), bundle 34, list 14, file 10: C.A. Bentley's note on 15 February 1921.

24. NAB, Report by S.N. Sur, Assistant Director of Public Health, Malaria Research Unit, Bengal, 24 September 1926. The debate as to whether mosquitoes could thrive with the water hyacinth was also an issue in the USA about this time. In response to the suggestion by a biologist that the growth of water hyacinth on the surface water was immediately followed by the destruction of mosquito larvae, another biologist, Alfred Weed, referring to the result of his own research, noted that the growth of water hyacinth over the water was 'followed by a great increase in the mosquito population'. See Alfred C. Weed, 'Another Factor in Mosquito Control', *Ecology*, 5 (January 1924): 110–11.

25. Banerjee, 'Some Economic Problems of Bengal – 1', 33.

26. CSAS, J.T. Donovan Papers, file I: Bengal 1927–1931: 'Tour Diaries of the Collector of Bakarganj'.

27. Babu Nibaran Chandra Das Gupta, in BLCP, 1, 3 (1921), 76–7; for other references to the devastation caused by the water hyacinth in Bengal see BLAP LIX, 4, 159; LIV, 2, 138; XXXIV, 3, 632. See also M. Azizul Huque, *The Man Behind the Plough* (Calcutta, 1939), p. 16; L.S.S. O'Malley, *Bengal District Gazetteers: Faridpur* (Calcutta, 1925), p. 8.

28. RWHC, 2, xv–xvii.

29. NAB, Revenue dept, 2 September 1915, bundle 34, list 14, file 202: J.R. Blackwood, Director of Agriculture, Bengal to Secretary to GoB,.

30. NAB, Progs nos. 24–27, July 1919, file 9-M-1: 'Water hyacinth', Communiqué of the Government of India, Revenue dept (Agriculture Branch) (hereafter 'Communiqué').

31. Robert S. Finlow, 'Water Hyacinth (kachuri, Tagoi, or Bilati Pana)', 9 November 1917, in 'Communiqué'.

32. From its birth in 1886, the Shaw Wallace Company dealt in a variety of different businesses including tea, Bengal silk, oil, tinplates, shipping of jute and gunnies, flour mills, coal mining, Swiss dyes and chemicals, and from 1944 to 1947 as one of the Bengal government's chief rice and paddy procurement agencies. In 1914, the company started operations in the business of fertilizer 'specializing in organic and inorganic mixtures for plantation and ryot crops'. It is not clear whether Shaw Wallace had a hand in introducing the water hyacinth into East Bengal or whether the water hyacinth drew the company to East Bengal. But it is probable that the company's turning to the Bengal water hyacinth may have been informed by the introduction of the Water Hyacinth Act in Burma in 1917, signalling restrictions in the potash trade. For a profile of the Shaw Wallace Company see 'The Varied Activities of Shaw Wallace', in the Bengal Chambers of Commerce Centenary Supplement, *ABP*, 24 March 1946, 26.

33. Messrs Shaw Wallace & Co., Calcutta to Fibre Expert to GoB, 12 August 1916, Revenue Department (Agriculture Branch), in 'Communiqué'.

34. Notice no. 2, in 'Communiqué'.

35. NAB, AID (Agriculture Branch), 'A' Proceedings nos. 26–27, bundle 34, list 14, file 9-M-11 (5): M.C. McAlpin, Secretary to GoB, Revenue dept, to All Commissioners of Divisions, 3 July 1919.

36. RWHC, 2.

37. Chaudhuri 'Agrarian Problems of Bengal', 118; see also Banerjee, 'Some Economic Problems of Bengal – 1', 34.
38. B.K. Banerjee, 'Water Hyacinth', *ABP*, 25 September 1935, 8. It may be noted that this argument was taken up in spite of the fact that Jagadish Chandra Bose himself objected to any chemical option as he feared that it could lead to water pollution affecting both human and fish.
39. Banerjee, 'Some Economic Problems of Bengal – 1', 34–5.
40. NAB, 'A' Proceedings for July 1919, nos. 24–27, file 9-M–11: 'Extracts from notes and orders in Agriculture', Revenue dept (Agriculture Branch). The metaphor of the 'Rabbits of Australia' appears pertinent from the following fact: 'Sometime in the 1850s a man was charged at the Colac (Victoria) Police Court with having shot a rabbit at the property of John Robertson of Glen Alvie. He was fined 10 pounds. A few years later, Robertson's son spent 5000 pounds a year in an attempt to control rabbits.' By 1869 it was estimated that 2,033,000 rabbits had been killed on his property and that they were as thick as ever. See 'Rabbits of Australia', http://web.archive.org/web/200001-200502re_/http://rubens.anu.edu.au/student.projects/rabbits/history.html, last accessed 11 November 2009.
41. NAB, AID (Agriculture Branch), bundle 34, list 14, file 6, 1921; see also NAB, AID (Agriculture Branch), bundle 34, list 14, file b155: G.P. Hogg, Secretary to GoB, to the Secretary to Govt of India, Dept of Education, Health and Lands, 3 August 1929.
42. *RWHC*, XXIX–XXX.
43. *RWHC*, III, VI.
44. NAB, AID (Agriculture Branch), bundle 34, list 14, file b155: G.P. Hogg, Secretary to GoB, to the Secretary to Government of India, Dept of Education, Health and Lands, 3 August 1929.
45. NAB, AID (Agriculture Branch), bundle 34, list 14, file 155: A.B. Reid, Joint Secretary to Govt of India, Dept of Education, Health and Lands, to Secretary to GoB, AID, 22 May 1929.
46. NAB, Public Health Department, 'B' Proceedings, nos. 110–115, list 14, bundle 15, file P.H. 2C-2: (Second) Report showing the progress made in giving effect to the recommendations of the Royal Commission on Agriculture in India. Part I – Central Government, for the period 1 November 1929 to 31 December 1930.
47. Bengal Act XII of 1934: The Bengal Waterways Act, 1934, section 41 (5); 'navigable channel' meant 'any channel which is navigable during the whole or a part of the year by a vessel of two-foot draught or over'. This approach to selective means to tackle the weed came at a time when all discussions of getting rid of the plant seemed to end in 'nothing but a tacit admission that it was too costly and altogether too vast an undertaking'. See CSAS, E.W. Holland Papers, 'Personal reminiscence', p. 32.
48. Bengal Act XIII of 1936: The Bengal Water Hyacinth Act 1936 and the Rules Thereunder, Clause 12.
49. Ray, *History of Public Health*, p. 295.
50. Bengal Act XIII of 1936.
51. Answer by Nawab K.G.M Faroqui to the question by Maharaja Giris Chandra Nandi, BLCP XLV, 342.
52. *Bengal Legislative Debates* (1939), 970–2.

53. *ABP*, 5 May 1939, 8.
54. *ABP*, 4 May 1939, 4.
55. Most of the English civil servants apparently enjoyed the occasional drives against the water hyacinth as these provided a break from routine office work. O.M. Martin recollected that 'in order to encourage this movement, my wife and I would don bathing costumes and do manual labour along with the villagers. Some of the villagers took up the work with great enthusiasm, and I could easily have made every villager join in the work, if I had allowed the vigorous minority to coerce the lazy majority. Sometimes the volunteers begged me for permission to take sticks in their hands, and drive the lazy ones to the work. I did not dare to give this permission, without any legal authority to do so; but at times I was sorely tempted to let the enthusiasts have their way. It was not a job that everybody liked, as the water-hyacinth was full of curious repulsive-looking creatures like snakes and crabs. But it was a healthy change from scribbling notes on office files.' See CSAS, O.M. Martin Papers: 'Memoirs of O.M. Martin, part II', p. 308.
56. Sudhir Chundar Sur, *ABP*, 14 April 1939, 16.
57. BLAP LIV (1939), 87–8.
58. BLAP LII (1938), 378–80.
59. NAB, Faridpur Files, General dept (Revenue), Collection no. 3A, File 20: K.C. Basak, Secretary to GoB, AID, to Collector, Faridpur, 7 June 1941.
60. NAB, Faridpur Files, General dept (Revenue), Collection no. 3A, File 20: Special Officer, Water-Hyacinth, to District Magistrate, Faridpur, 17 February 1941.
61. 'Attempts to Destroy the Water Hyacinth,' *Dacca Prakash*, 3 March 1935, 5.

8.   Between Food Availability Decline and Entitlement Exchange: an Ecological Prehistory of the Great Bengal Famine of 1943

1. Sister Nivedita (Margaret Elizabeth Noble) was an Irish disciple of Swami Vivekananda, the great Vedanta philosopher of India. She arrived in India in 1898 and lived and worked there until her death in 1909.
2. Sister Nivedita, *Glimpses of Famine and Flood in East Bengal in 1906* (Calcutta, 1907), p. 28.
3. Nivedita, *Glimpses of Famine*, p. 54.
4. Mahalanobis, the great Indian statistician, remarked that the famine of 1943 was not an accident like an earthquake or a flood, but the 'culmination of economic changes which were going on even in normal times'. See P.C. Mahalanobis, 'The Bengal Famine: the Background and Basic Facts', *Asiatic Review*, XLII (January 1946): 315; see also APAC, Mss Eur 911/8, Pinnell Papers: 'Confidential Memorandum on the Economic Condition of Bengal Prior to the Famine of 1943', pp. 294–5.
5. Tim Dyson and Arup Maharatna, 'Excess Mortality during the Bengal Famine: a Re-evaluation', *Indian Economic Social History Review*, 28 (1991): 283.
6. The concept of 'entitlement exchange' is succinctly described by Amartya Sen: 'Starvation is the characteristic of some people not *having* enough

food to eat. It is not the characteristic of there *being* not enough food to eat.' Amartya Kumar Sen, *Poverty and Famines: an Essay on Entitlement and Deprivation* (Oxford, 1981, reprinted 1988), p. 1.

7. Utsa Patnaik, 'Food Availability and Famine: a Longer View', *Journal of Peasant Studies*, 19(1) (1991): 1–12; David Arnold, *Famine. Social Crisis and Historical Change* (Oxford, 1988). For a summary of the debates around Sen's approach and criticisms of it, see S.R. Osmani, 'The Entitlement Approach to Famine: an Assessment', Working Paper no. 107, UNU World Institute for Economic Research, Helsinki, 1993.

8. The food availability decline (FAD) approach explains famine in terms of the reduction of food production or supply.

9. For perspectives on FAD as instrumented by natural disasters, see Mark B. Tauger, 'Entitlement, Shortage and the 1943 Bengal Famine: Another Look', *Journal of Peasant Studies*, 31(1) (October 2003): 47–8. One of the few exceptions that take a long-term ecological context is Vinita Damodraran, 'Famine in Bengal: a Comparison of the 1770 Famine in Bengal and the 1897 Famine in Chotanagpur', *Medieval History Journal*, 10(1–2) (2007): 143–81.

10. Cheng Siok-Hwa, *The Rice Industry of Burma 1852–1940* (Kuala Lumpur, 1968), p. 211.

11. S. Bose, *Peasant Labour and Colonial Capital: Rural Bengal since 1770* (Cambridge 1993), p. 23.

12. Arthur Geddes, 'The Population of Bengal, its Distribution and Changes: a Contribution to Geographical Method', *Geographical Journal*, 89(4) (April 1937): 344.

13. M. Mufakharul Islam, *Bengal Agriculture 1920–1946* (Cambridge, 1978), p. 200; for two earlier but still relevant studies of decline or stagnation in India's agriculture and per capita food output in the first half of the twentieth century, see Daniel Thorner and Alice Thorner, *Land and Labour in India* (Bombay, 1965) and George Blyn, *Agricultural Trend in India* (Philadelphia, 1966).

14. Islam, *Bengal Agriculture*, pp. 201–2.

15. CSAS, Martin Paper, part II, p. 250.

16. H.S.M. Ishaque, *Agricultural Statistics by Plot to Plot Enumeration in Bengal 1944 and 1945*, (part II, Alipore, 1947), pp. 5, 15–17, 21, 47–9, 69–70, 73, 75, 77–8, 82–4.

17. Ishaque, *Agricultural Statistics*, p.6.

18. APAC, Mss Eur D911/8: Pinnell Papers, pp. 6, 23.

19. *ABP*, 20 Sept 1942, p. 6.

20. *Annual Report of the Department of Agriculture, Bengal, 1929–30* (Calcutta, 1930), p. 19. (Hereafter ARDA.)

21. Ibid., p. 19.

22. See ARDA, for the years 1933–34, part 2, p. 9; 1934–35 part 1, p. 7; 1935–36, part 2, p. 8; 1936–37, part 2, pp. 26–7; 1938–39, part 1, p. 11; 1938–39, part 2, p. 3; 1939–40, part 2, p. 15; 'Ufra disease of rice', Department of Agriculture, Bengal: Bulletin no. 2 (Calcutta, 1912), pp. 2–3.

23. S.Y. Padmanabhan, 'The Great Bengal Famine', *Annual Review of Phytopathology*, 11 (1973): 23. Other factors promoting this disease include lack of nutrient elements in the soil, accumulation of toxic elements in the soil, humidity and wet conditions.

24. A form of this disease ravaged the deltaic tracts of Krishna and Godaveri in 1918–19. It was reported that unusually heavy rains at the time of ear formation of the plants followed by flooding and ill-drained conditions had caused this disease. Padmanabhan, 'The Great Bengal Famine', 12.
25. 'Brown Spot' International Rice Research Institute, Manila, http://www.knowledgebank.irri.org/ricedoctor/index.php?option=com_content&view=article&id=557&Itemid=2762, last accessed 19 December 2009; for a recent discussion on the effect of 'brown spot' diseases of rice and the Bengal famine, see Tauger, 'Entitlement, Shortage', pp. 63–6.
26. Padmanabhan, 'The Great Bengal Famine', p. 11.
27. APAC, MSS EUR F125/163, 20 March 1943.
28. *Report of the Foodgrains Procurement Committee* (Calcutta, 1944), p. 4.
29. Memorandum of the Hon'ble Mr Justice H.B.L Braund on events from March 1943 to the end of 1943 in relation to the food situation in Bengal (Calcutta), p. 7.
30. M. Mufakharul Islam, 'The Great Bengal Famine and the Question of FAD Yet Again', *Modern Asian Studies*, 21(2) (2007): 437.
31. For those without meaningful entitlement to land, market made little sense. For a broader argument in this respect, see Stephen Devereux, 'Sen's Entitlement Approach. Critique and Counter-critiques', *Oxford Development Studies*, 29(2) (2001).
32. See for instance Patnaik, 'Food Availability and Famine'; Islam, 'The Great Bengal Famine'.
33. LRC, 1 (1934), 37.
34. Comilla Collectorate Paper, case number 10855 of 1920, in the court of the HSO of Tippera; Comilla Collectorate Paper, Plaintiff: Mohanta Bhagaban Das, Defendant: Nasaruddin and others, Case no 10860 of 1920; District Dhaka settlement court, ashuganj camp, Plaintiff: Kumar Kamalranjan Rai, Defendant: Bhagban namasudra, Dacca Settlement ASO Case number 1474 of 1917.
35. Mohammed Mohsen Ullah, *Burir Suta* (Rajshahi, Calcutta: 1317 BS; AD 1911), pp. 27, 28.
36. Tripura Settlement Office, Chandua camp, document no. 10859 of 1920.
37. NAB, GoB, Revenue dept (Land), June 1927, progs no. 7–10, wooden bundle no. 76, list 17: P.C. Mitter, 'Notes regarding the Bengal Tenancy Amendement Bill, 12 December 1923.
38. David Ludden, 'Introduction: Agricultural Production and Indian History', in David Ludden (ed.), *Agricultural Production and Indian History* (Oxford, 1994), p. 4.
39. Faridpur Collectorate, District Record Room, Tauzi no. 6506: Reajuddin Sarkar to the District Settlement Officer, Faridpur, dated 30/3/1913.
40. NAB, Progs of the cooperative credit and rural indebtedness (CCRI) dept (Rural Indebtedness), file 2M/40, 'B' progs, bundle 1, list 36: E.W. Holland, Office of the Director, Debt Conciliation, Western Circle (p. 2), 12 June 1937; *BLCP*, 48th session, 1936, vol. XLVIII, no. 2, p. 17.
41. Floud Commission Report, vol. IV, p. 54; Birendra Kishore Roychowdhury, *Permanent Settlement and After* (Calcutta, 1942), p. 292.
42. Sen, *Poverty and Famines*, pp. 70–5.
43. Chatterjee, *The Present History of West Bengal* (Delhi 1997), pp. 61–4.

44. Nivedita, *Glimpses of Famine*, pp. 49, 62.
45. Antonio Gramsci, *Selections From the Prison Notebooks*, ed. and trans. by Q. Hoare and G.N. Smith (London, 1971), p. 213.
46. This argument is heavily drawn from CSAS, (John) Bell Papers, box V, no. 37: 'The Bengal Famine. 1942–43', p. 25; On land-holders' preference to pay in cash instead of kind during the rise of rice prices, see LRC, 6 (1941), 46.
47. Sharecropping in Bengal was 'a stage in between landholding and landlessness'. See A. Cooper, *Sharecropping and Sharecroppers' Struggles in Bengal 1930–1950* (Calcutta, 1988), p. 91.
48. P.C. Mahalanobis, 'The Bengal Famine', p. 313.
49. Ibid.
50. *The Times*, 3 November 1943: 5; for the bhadralok's relative immunity from famine, see Indivar Kamtekar, 'A Different War Dance: State and Class in India 1939–1945', *Past and Present*, 176(1) (2002): 218 fn.
51. Tim Dyson, 'On the Demography of South Asian Famines, part II', *Population Studies*, 45(2) (July 1991): 279–97.
52. Arup Maharatna, 'Malaria Ecology, Relief Provision and Regional Variation Mortality during the Bengal Famine of 1943–44', *South Asia Research*, 13(1) (May 1993).
53. Churchill College, Cambridge, Churchill Archives, AMEL 2/3/19: Casey to Amery, Sec for State to India, 24 Feb 1944, p. 2.
54. *Annual Sanitary Report of the Province of Eastern Bengal and Assam*, 1905, pp. 1–10.
55. *Census of India (Bengal)*, 1901, part 1, p. 73.
56. C.A. Bentley, *Malaria and Agriculture in Bengal. How to Reduce Malaria in Bengal by Irrigation* (Calcutta, 1925).
57. *Annual Sanitary Report*, 1909, p. 13; for a note on relatively better health conditions in some places in nineteenth-century Bengal, see, Rai Chunilal Bose, *Palli-swashtha O saral swashtha-bidhana* [Village Sanitation and a Manual of Hygiene] (4th edn, Calcutta, 1934), p. 39.
58. Kiran Chandra De, *Report on the Fisheries of Eastern Bengal and Assam* (Shillong, 1910), pp. 71–3.
59. Ibid., p. 70.
60. Nivedita, *Glimpses of Famine and Flood*, pp. 62–3.
61. For a list of statements on earnings and expenditure on different sectors at both district and union levels, see *Resolution Reviewing the Reports on the Working of District and Local Boards in Bengal between 1932–33 and 1939–1940*.
62. 'Since the protein in fish supplement the growth-promoting effect of the pulse proteins, it is highly probable that the pulse proteins will be utilized more efficiently when they are included in the diet containing fish.' See K.P. Basu and H.N. De, 'Nutritional Investigation of Some Species of Bengal Fish', *Indian Journal of Medical Research*, 26(1) (July, 1938): 188. (Hereafter *IJMR*.)
63. A.B. Fry, 'Indigenous Fish and Mosquito Larvae: a Note from Bengal', *Paludism: Being the Transaction of the Committee for the Study of Malaria in India*, 1(5) (September 1912): 71–4.
64. K.P. Basu and M.N. Basak, 'Biochemical Investigations on Different Varieties of Bengal Rice – part V', *IJMR*, 24(4) (April 1937): 1067.

65. Ibid.
66. For a note on the shift to the boro variety because of the water hyacinth and other factors, see Ratan Lal Chakraborty, 'Reclamation Process in the Bengal Delta', in Yoshihiro Kaida (ed.), *The Imagescape of Six Great Asian Deltas in the 21st Century* (Kyoto, 2000), p. 26.
67. Prafullah Kumar Bose remarked in the early 1940s that the damage caused to crops and fishs by the water hyacinth alone was 'crores' (1 crore = 10 millions) of rupees. Quoted in 'Geographical Record', *Geographical Review*, 36(2) (April 1946): 229–30.
68. For instance, the need for community level awareness of cleanliness was taken up, and among other initiatives in the early twentieth century, 15,000 Azolla plants were sent to various parts of India as part of an experiment to prevent the growth of mosquito larvae in stagnant bodies of water. *Annual Report of the Royal Botanic Garden Calcutta*, 1909–10, p. 2.
69. David Arnold's *Colonizing the Body: State Medicine and Epidemic Disease in Nineteenth-Century India* (Berkeley, 1993) remains a classic in this area of investigations.
70. 'Extract from the Progs of the Lieutenant-Governor of Eastern Bengal and Assam in the Municipal dept, dated 11th October 1910', in *Paludism*, 1(2) (January 1911): 93–8.
71. David Arnold, *Science, Technology and Medicine in Colonial India* (Cambridge:, 2000), p. 57; see also Iftekhar Iqbal, 'Modernity and its Discontents. Studying Environmental History of Colonial and Postcolonial Bangladesh' *East West Journal*, 1(1) (2007).
72. In 1912, A.B. Fry of the Indian Medical Service suggested that although the creation of borrow pits was necessary in the construction of railways, roads or houses, the 'aim in water-logged areas should be to make them [borrow pits] large and deep so that they can never dry and kill off the fish'. In this way Fry felt that the spread of mosquito larvae could have been checked. He suggested that the railway companies should be made to appoint a full-time executive officer and staff to attend to the supervision of borrow pits and keep them in a sanitary condition. See A.B. Fry, 'Indigenous Fish and Mosquito Larvae', pp. 73–4. But, as we have seen, the railway authorities were not prepared to undertake such programmes.
73. For the devastation caused by malaria in the embanked coastal area, see M.O.T. Iyengar, 'The Distribution of *Anopheles Ludlowii* in Bengal and its Importance in Malarial Epidemiology', *IJMR*, 19(2) (October 1931): 499–524.
74. See *IJMR*, 21(4) (April 1934): 935. For a general focus on colonial and post-colonial India's relation with international health bodies, see Sunil Amrith, *Decolonizing International Health: India and Southeast Asia 1930–65* (Basingstoke: Palgrave Macmillan, 2006). For a note on the humane mission and public value of British imperial science and its agronomic failure to provide quinine at a price affordable to the public, see Richard Drayton, *Nature's Government: Science, Imperial Britain, and the 'Improvement' of the World* (New Haven and London, 2000), p. 231.
75. 'Factsheets on Chemical and Biological Warfare Agents', http://www.cbwinfo.com/Biological/Pathogens/VC.html, last accessed on 15 November 2009.
76. 'The "pot well" was an earthen shaft lined by locally burnt pot rings with a diameter of 2 feet 9 inches. Such wells are sunk usually in the dry season,

and are carried down to a depth of between 20 and 30 feet, i.e., until a depth of water of 4 to 6 feet is obtained. The well lining is generally built up some 3 feet above the ground surface and supported by a small earth ramp.' T.H. Bishop, 'The Working of the Cholera Prevention Scheme on the Lower Ganges Bridge Construction', *IJMR*, 1(2) (1913–14): 300.
77. For instance, a member of the Bengal Legislative Council alleged that cholera was particularly severe along the railways. See BLCP, 5th session, 1921, vol. 5, p. 500.
78. Bishop, 'The Working of the Cholera Prevention Scheme', p. 297.
79. Bishop noted that the average annual costs of the sinking and repair or maintenance was Rs.185 and Rs. 330 for shallow and deep tube wells respectively (ibid., pp. 299–300); Chunilal Bose estimated it to be 70 to 80 taka and 5 to 6 thousand taka respectively, Bose, *Palli-swastha*, p. 74. In any case these amounts were prohibitive for the poor who could not earn that sum of money in a year.
80. Ira Klein, 'Death in India, 1871–1921', *Journal of Asian Studies*, 32(4) (August 1973): 659.
81. In 1943, the number of deaths in Bengal increased by 58 per cent on the 5-year average. Where the average annual number of deaths was 1,184,903 in normal years, in 1943 it rose to 1,873,749. In 1943, the number of deaths due to cholera was 160,909 more than the normal annual average. In the same year, the number of deaths due to malaria was 285,792, more than the normal annual average. See 'Banglar Mrittu Sankya'['An account of the mortality rate in Bengal'], *Dacca Prakash*, 12 March 1944, p. 2. For a note on the sharp increase of diseases in Bengal since the late 1930s, see K.S. Fitch, *A Medical History of the Bengal Famine 1943–44* (Calcutta, 1947), p. 124.

9. Reflections

1. For the pioneering account of the 'world system', see Emmanuel Wallerstein, *The Modern World-System* (New York, 1974).
2. 'X', Quoted in *Eastern Bengal and Assam Era*, 14 February 1906.
3. John L. Christian, 'Anglo-French Rivalry in Southeast Asia: its Historical Geography and Diplomatic Climate', *Geographical Review*, 31(2) (1941): 278.
4. Arthur Cotton, 'On Communication Between India and China by the Line of the Burhampooter, June 24, 1867', *Proceedings of the Royal Geographical Society of London*, 11(6) (1866–67): 257.
5. John Ogilvy Hay, 'A Map Shewing the Various Routes and Yang-tse Connecting China with India and Europe Through Burmah and Developing the Trade of Eastern Bengal, Burmah & China' (London: Edward Stanford, 1875).
6. G.N. Gupta, *A Survey of the Industries and Resources of Eastern Bengal and Assam for 1907–1908* (Shillong, 1908), p. 102.
7. Ibid., p. 105.
8. David Ludden, *Imperial Modernity and the Spatial History of Global Inequality* (forthcoming, 2011).
9. This was as much a global process. Note Michael Williams's observations: 'the 1880s was a period of particular introspection. For Europe the space for colonization had all but gone, and the glitter and brilliance of "la belle

epoch" seemed too brittle to last. For the US a similar sense of limited space accompanied Frederick Jackson Turner's announcement that the frontier had "closed" and the Gilded Age was over. The phrase fin de siècle was coined; more than a reference to the last decade, it resonated: decade, decayed, decadence.' Michael Williams, *Deforesting the Earth: From Prehistory to Global Crisis, An Abridgement* (Chicago, 2006), p. 359.

10. See 'Preface' to David Ludden (ed.), *Agricultural Production, South Asian History, and Development Studies* (New Delhi, 2005).

11. For the debates, see Joya Chatterji, *Bengal Divided: Hindu Communalism and Partition, 1932–1947* (Cambridge, 1994), and P. Chatterjee, 'On Religious and Linguistic Nationalisms: the Second Partition of Bengal', in Peter van der Veer and Hartmut Lehmann (eds), *Nation and Religion: Perspectives on Europe and Asia* (Princeton, 1999), pp. 116–26.

12. For details of the 'surgical' fashion in which East Bengal's map was drawn during partition see, Joya Chatterji, 'The Fashioning of a Frontier: the Radcliffe Line and Bengal's Border Landscape 1947–52', *Modern Asian Studies*, 33(1) (1999): 185–242.

13. David Arnold, *Science, Technology and Medicine in Colonial India* (Cambridge, 2000), p. 209.

14. For a recent collection of important essays that examine aspects of the relationship between nationalism and ecology in South Asia, see Gunnel Cederlöf and K. Sivaramakrishnan (eds), *Ecological Nationalisms: Nature, Livelihoods, and Identities in South Asia* (Seattle, 2006).

15. Srischandra Nandy, 'Preface', *Bengal Rivers and our Economic Welfare* (Calcutta, 1948), pp. 59–63.

16. Uttam Kumar Chowdhury et al., 'Groundwater Arsenic Contamination in Bangladesh and West Bengal, India', *Environmental Health Perspective* (May 2000), http://findarticles.com/p/articles/mi_m0CYP/is_5_108/ai_63322036?tag=artBody;col1, last accessed 15 November 2009.

17. For an overview of the project and its implications, see Yoginder K. Alagh et al. (eds), *Interlinking of Rivers in India: Overview and Ken-Betwa link* (New Delhi, 2006). For concern about sustainability and the colossal ecological impact of the project see S.R. Singh and M.P. Srivastava (eds), *River Interlinking in India: the Dream and Reality* (New Delhi, 2006); S.M. Sengupta, 'Interlinking of Rivers May Cause Geomorphic Changes', *Journal of Geological Society of India*, 69(5) (2007): 11–34; A.K. Saxena Misra, A. Yaduvanshi, M. Mishra et al., 'Proposed River-Linking Project in India: Boon or Bane to Nature?', *Environmental Geology*, 51(8) (2007): 1361–76; A.R.M. Khalid, 'The Interlinking of Rivers Project in India and International Water Law: an Overview', *Chinese Journal of International Law*, 3(2) (2004); Medha Patkar (ed.), *A Millennium Folly?* (National Alliance for People's Movement and Initiatives, 2004), pp. 553–70; J. Bandyopadhyay and S. Perveen, 'Interlinking of Rivers in India: Assessing the Justifications', *Economic and Political Weekly*, 39(50) (2004): 5307–16; J. Bandyopadhyay and S. Perveen, 'Doubts over the Scientific Validity of the Justifications for the Proposed Interlinking of Rivers in India', *Science and Culture*, 70(1–2) (2004): 7–20.

18. Cederlöf and Sivaramakrishnan, *Ecological Nationalisms* and Rohan D'Souza, *Drowned and Dammed : Colonial Capitalism, and Flood Control in Eastern India* (Delhi: Oxford University Press, 2006), pp. 20–45, 223–4. For a note on the

recent assertion of ecological nationalism on the part of China, India and Bangladesh in terms of the eastern Himalayas water systems, see Iftekhar Iqbal, 'Making Sense of Water', *Forum*, 2(5) (June, 2007).

19. For the debates see Richard Palmer-Jones, 'Slowdown in Agricultural Growth in Bangladesh: Neither a Good Description Nor a Description Good to Give', pp. 92–136 and Shapan Adnan, 'Agrarian Structure and Agricultural Growth Trends in Bangladesh: the Political Economy of Technological Change and Policy Interventions', pp. 177–228, in Ben Rogaly, Barbara Harriss-White and Sugata Bose (eds), *Sonar Bangla? Agricultural Growth and Agrarian Change in West Bengal and Bangladesh* (New Delhi, 1999).

20. Ben Crow, 'Why is Agricultural Growth Uneven? Class and the Agrarian Surplus in Bangladesh', in Rogaly et al., *Sonar Bangla?* pp. 147–76.

21. An important narrative of modernist displacements of the water regime in colonial and postcolonial Bangladesh is Ahmed Kamal, 'Living with Water. Bangladesh since Ancient Times', in T. Tvedt and E. Jakobsson (eds), *A History of Water*, vol. 3 (London, 2006), pp. 200–10.

22. Nusha Yamina Choudhury, Alak Paul and Bimal Kanti Paul, 'Impact of Coastal Embankment on the Flash Flood in Bangladesh', *Applied Geography*, 24(3) (July 2004): 241–58.

23. Shapan Adnan, 'Intellectual Critiques, People's Resistance, and Inter-Riparian Contestations: Constraints to the Power of the State Regarding Flood Control and Water Management in the Ganges-Brahmaputra-Meghna Delta of Bangladesh', in Devleena Ghosh, Heather Goodall and Stephanie Hemelryk Donald (eds), *Water, Sovereignty, and Borders in Asia and Oceania* (London, 2009), pp. 110–17.

24. *The Lancet*, 359(9312), 30 March 2002: 1127.

25. See 'Boimela O Nari-samabesh: the First National Feminist Book Fair – 1999', http://membres.lycos.fr/ubinig/eventboimela.htm, last accessed 15 November 2009.

26. Abul Barkat, Shafique uz Zaman and Selim Raihan (eds), *Political Economy of Khas Land in Bangladesh* (Dhaka, 2001), p. 86. A more recent report put this figure at 5 million acres of khas land (8.7 per cent of total land), Towheed Feroze, 'Land Policy: Pro-Poor Plan is the Key', *Dhaka Courier*, 6–12 June, 24(46) (2008): 16–17.

27. It is predicted that about 8 per cent of the territory of Bangladesh is likely to be submerged by the rising sea by 2050 due to climate change. But Bangladesh is, at the same time, one of the few countries where a dynamic geological process of land-formation is going on in the coastal region. Recent GIS images and data interpolated since 1973 suggest that the landmass of Bangladesh has actually increased by 1000 square kilometres and it is expected that it can gain 1000 sq km more of land by 2050. This, if scientifically proved, speaks of even better prospect for the landless in Bangladesh. For the reports on formation of new lands in Bangladesh, see 'Bangladesh "is Growing"', *Timesonline*, 1 August 2008, http://www.timesonline.co.uk/tol/news/environment/article4440982.ece, last accessed 23 November 2009.

28. For a recent example of illegal encroachment on char lands, see *Daily Star*, 20 April 2008; for a narrative of using state mechanism for appropriating the khas mahals in Chittagong Hill Tracts and char lands, see Shapan Adnan, *Bangladesher Krishi Prashno: Bhumi Sanskar O Khas Jamir Odhikar Protishthai*

Gonoandoloner Bhumika [The Agricultural Question in Bangladesh: the Role of Mass Movement in Land Reforms and the Establishment of Rights on Khas Lands] (Dhaka, 2008).

29. Peter Bertocci, 'Structural Fragmentation and Peasant Classes in Bangladesh', *Journal of Social Studies*, 5 (1979); for evidence of the centripetal mobility of peasant society in the 1970s, see Shapan Adnan and H. Zillur Rahman, 'Peasant Classes and Land Mobility: Structural Reproduction and Change in Rural Bangladesh', *Bangladesh Historical Studies*, 3 (1978); Abu Abdullah, Mosharaff Hossain and Richard Nations, 'Agrarian Structure and the IRDP – Preliminary Considerations', *Bangladesh Development Studies*, 4(2) (1976); Rogaly et al., 'Introduction', *Sonar Bangla?* p.18; also Adnan, 'Agrarian Structure and Agricultural Growth Trends'.

30. For the details of these arguments, see Mushtaq Husain Khan, 'Power, Property Rights and the Issue of Land Reform: a General Case Illustrated with Reference to Bangladesh', *Journal of Agrarian Change*, 4(1–2) (2004). This argument is illustrated by the fact that a few weeks after the formation of a new democratic government following what has been termed as one of the most successful national elections in Bangladesh, six people were killed in clashes between the factions of different political parties over the possession of a huge char on the Padma.

Appendixes

Appendix 1: specimens of forms of pottah (land settlement contract) in the 1870s

Appendix 1a: form of pottah for a cultivator having a right of possession at fixed rates

This is the pottah granted to ____, a ryot of mouzah ____ in pergunnah ____, holding a permanent right at a fixed rent under the following conditions:—

In accordance with the results arrived at in the present settlement of the aforesaid mehal, your holding consists of ____ bighas (equal to ____ acres) of land of the several descriptions enumerated below. The jumma of the aforesaid holding has been ascertained to be, and is hereby confirmed at Rs. ____. You bind yourself to pay this jumma year by year to the person entitled to receive the rents of the mehal, according to the instalments specified at foot, and in return you will obtain a receipt for every sum paid as rent to such person as aforesaid.

You possess a right of permanent occupation in the lands now held by you at the rent specified in this pottah. No one has the power to evict you from such lands at will, or to enhance your rent on account of such lands so long as you duly pay your rent. You are entitled to take and enjoy the fruits of existing trees, or of such trees as may be planted by you in future within the area of your present holding. You have also a right to cut or sell every tree planted and reared by yourself or by your ancestors on your present holding. The rights, declared by this pottah to vest in you will devolve on your heirs at your death.

Appendix 1b: form of pottah for a cultivator having a right of occupancy

This is the pottah granted to ____, a ryot of mouzah ____ in pergunnah ____, holding a permanent right at a fixed rent under the following conditions:—

In accordance with the results arrived at in the present settlement of the aforesaid mehal, your holding consists of ____ bighas (equal to ____ acres) of land of the several descriptions enumerated below. The jumma of the aforesaid holding has been ascertained to be, and is hereby confirmed at Rs. ____. You bind yourself to pay this jumma year by year to the person entitled to receive the rents of the mehal, according to the instalments specified at foot, and in return you will obtain a receipt for every sum paid as rent to such person as aforesaid.

You possess a right of occupancy in the lands now held by you. Nether the Government nor the person with whom a settlement of the mehal may have been made has power to evict you from such lands at will, nor to enhance your rent on account of such lands except in accordance with the provisions of the law relating to such enhancement. You are entitled to take and enjoy the fruits of existing trees, or of such trees as may be planted by you in future within the area of your present holding. The rights declared by this pottah to vest in you will devolve on your heirs at your death.

Appendix 1c: form of pottah for a cultivator not known to possess a right of occupancy*

This is the pottah granted to ____, a ryot of mouzah ____ in pergunnah ____, under the following conditions:—

In accordance with the results arrived at in the present settlement of the aforesaid mehal, your holding consists of ____ bighas (equal to ____ acres) of land of the several descriptions enumerated below. The jumma of the aforesaid holding has been fixed for the present at Rs. ____. You bind yourself to pay this jumma year by year till further notice according to the instalments specified at foot.

* It is explained that, 'But if it be claimed, or if claimed it to be not proved, a pottah must be given in Form III will properly leave the question altogether untouched, and at the same time will not bar the right accruing should it, though non-existent at the time when the pottah is given, legally accrue subsequently.'

Source: NAB, Revenue (land), 'A' progs, wooden bundle 13, list 17: S.S. Collection I.—no.14, August 1873 (General).

Appendix 2: specimen form of pottah in the early twentieth century (for a bargadar)

I have prayed for cultivating the same [a certain amount of land] in partnership (bhage), you have granted my prayer ... I hereby pledge that I shall cultivate only within the specified portion of the land. I shall not do anything, as digging or any other things, that would decrease the fertility or value of the land. I shall always grow paddy on the land. Every year, before growing crops, I shall ask you what kind of paddy shall have to be grown on which land, and shall grow paddy according to your desire. I shall not be able to grow any paddy or crops according to my own will. I will not cultivate any variety of paddy of my own choice or any other crop. When in due course the paddy would be ripe, I shall cut it and take it to your house for thrashing ... I shall get the remaining paddy (after delivering a certain quantity of paddy to the owner) and hay as remuneration for my cultivation, seeds, looking after and labour instead of money in cash. If I do not deliver the crop in due course of time you are free to claim it by legal or informal means with demurrage and at a market price; I will have no objection to it. If you ever desire to produce crops by yourselves, then I will leave the land without any excuse and will not claim any right on the land.

Source: Rev dept (land), progs of GoB, Jan 1922, pp. 133–4.

Appendix 3: a comparative statement of the number of khas mahals to settle during 1849–50 in greater Bengal

Division	District	(a) Number of government mahals capable of immediate settlement	(b) Number of resumed mahals capable of immediate settlement
Jessore	Bancoorah	0	1
	Barasat[a]	0	0
	Burdwan	17	42
	Hooghly	3	17
	Nadia	21	48
	24-Parganah	15	31
	Jessore[b]	28	17
Total		84	156

(*continued*)

Appendix 3: Continued

Division	District	(a) Number of government mahals capable of immediate settlement	(b) Number of resumed mahals capable of immediate settlement
Murshidabad	Birbhoom	0	0
	Bogra	0	0
	Murshidabad	5	0
	Pubna[b]	3	3
	Rajshahi[b]	9	2
	Rangpur	4	7
Total		21	12
Dhaka	Bakarganj[b]	13	4
	Dhaka[b]	12	59
	Faridpur[b]	18	9
	Mymensingh[b]	33	64
	Sylhet[a]	43	80
Total		119	216
Chittagong	Bullooah[b]	40	35
	Chittagong[b]	159	44
	Tipperah[b]	37	13
Total		236	92
Patna	Behar	22	42
	Patna	7	18
	Sarun	0	22
	Shahabad	2	5
Total		31	87
Bhagalpur	Bhagalpur	10	34
	Danajpur	1	4
	Maldah	2	2
	Monghyr	7	20
	Purnea	15	135
	Tirhoot	2	12
Total		37	207
Cuttack	Balasore	0	1
	Cuttack	0	4
	Khoordah	0	0
	Midnapur	1	5
Total		1	10

Appendix 3: Continued

Division	District	(a) Number of government mahals capable of immediate settlement	(b) Number of resumed mahals capable of immediate settlement
Grand total	35 (number of districts)	529	780
Total for deltaic districts	11 (number of districts)	429	330

Notes:
(a) Represents both active and moribund deltaic characteristics.
(b) Active delta district.
Source: Report on the Revenue Administration of Lower Provinces, 1849–50, p. 35 (IOR V series, v/24/2497).

Appendix 4: number of Hindu weaving caste and of those who were actually engaged in weaving in 1870

	24 Parganas	Nuddea	Jessore	Pabna	Dacca	Faridpur	Bakarganj	Noakhali	Tippera
Weaving castes	70,203	37,834	52,485	21,719	42,528	14,481	38,627	34,733	76,801
Actually engaged in weaving	6,120	13,680	20,009	15,684	17,876	14,723	14,146	28,492	19,804

Source: Report on the Census of Bengal, 1871 (Calcutta, 1872).

Appendix 5: a petition from some Faraizis

When the Charghat estate, consisting of Mouza Nuj Charghat, Rosae Sohooah, Kolatubar, and seven other Mouzahs in Pergunnah Ookrah, was purchased by Rammohun Banerjee, zemindar, he commenced measuring our Maurossee lands, with the view of raising the rents. But, on our ancestors objecting, the measurement was put to a stop to by an order of the judge of Nuddea. Subsequently, when Baboo Eshwar Chunder Moostophee purchased this mehal, and during the time it was held on putnee tenure by Kylasanth Bromocharee, we were left in undisturbed possession of our lands.

To our great misfortune, about two years ago, Mr. Larmour, manager of the Moelatee and other factories, took a lease of these villages, and demanded from us the expenses which the izarah had entailed upon him, as well as other cesses. As we refused to pay, he at once proceeded to measure our land, which he had no business to do. In this, however, he has been stopped, by our petitioning the court. Still, in consequence of the enmity which he has on this account conceived against us, our lives, honour and property have been with great difficulty preserved during the last few months. As an instance of his oppression, we may mention that we had sowed our lands with sugar-cane and tobacco, and since congratulating ourselves on the fair prospects of our crop, when, on the 26th of Falgoon last, Dewan Nobin Chunder Chowdry, of the Barrorea factory of Mr. Larmour, Kazimuddy Ameen, Muloo Kholassee, and Jameeroodee Kholassee, accompanied by 50 or 60 lattials, armed with clubs, spears, and guns, and a Keranee Saheb on horseback, came to our villages, and having forcibly seized us, like judgement debtors, vowed vengeance to us and to our families, if we did not immediately sow our lands with indigo, and by forcibly bringing out the ploughs of some of the ryots from their houses, and by maltreating others by putting the ploughs on their necks, they got some of the lands cultivated with indigo. These people now come daily to our villages, and with similar parade practise oppressions. The oppressions of the indigo planters are beyond description. They have been mentioned in the 'Englishman,' 'Hurkaru,' and other papers, and several people have written on the subject, but no one has been able to fully describe them. Several officers of Government have devised, and we hear that the Government even now are devising, measures with a view to check these oppressions. But Mr. Larmour heeds them not. He has now adopted a plan by which he hopes to bring under his control some of the principal ryots of our turruff, by holding out to them hopes of employment. But, be this as it may, we, poor ryots, find no means for the protection of our honour, lives, and property. Police officers are posted for the prevention of the peace; yet even they dare not come forward at the critical time. We, therefore, pray that the huzoor will issue suitable orders to prevent the planters from forcibly sowing indigo in our lands, and to protect our lives and property. At the time we came to the huzoor, we left our families in the houses of our friends residing in the neighbouring villagers, and we do not know what may have happened in our own villages during our absence. We solicit that early orders may be passed for taking recognizances from the factory Dewan, Ameen, and Kholasees. We never took any indigo advances from Mr. Larmour, and even for the indigo, which, under the orders of the Huzoor, we cultivated last year, Mr. Larmour has not paid us a cowree. We pray, therefore, that suitable orders may be passed on our representation.

Note: Translation of a Petition from Muneroodeen Mundle and others, dated 1 Chyte, 1265 BS (AD 1859) to the Joint Magistrate of Baraset.

Source: *Report on the Indigo Commission*, p. 109.

Appendix 6: excerpts from a petition by a bhadralok

Dated Singardaha, the 19th January 1927

From – Babu Surja Kumar Guha Roy

To – The Collector of Faridpur

I fear I am intruding upon your precious time and memory when I present my rather forgotten self before you. May I remind you that I had the pleasure of acquainting myself with you when you visited the local Edilpur High School, where I am a member of the School Committee. Perhaps, it is in that connection that you knew me as the father of Mr. S.N. Guha Roy, I.C.S., now stationed at Brahmanbaria (Tippera).

Now to go to my point direct. Your evidence before the Agricultural Commission at Dacca emboldens me to approach you for a favour, which I hope, you will not fail, if circumstances allow, to show me. I hear that a good deal of khas mahal lands in Megnar Char and Padmar Char in the district of Faridpur is going to be settled this year. My son, D.N. Guha Roy, who has just got his M.A. degree of the Calcutta University is after taking to agriculture in right earnest, and if you be kind enough to give him scope in his native district, it will be extremely advantageous for him to begin his career. He is seeking to get the necessary preliminary training in some agricultural farm and if he is supplied with sufficient land, he is earnest after beginning agriculture in the improved methods employed in some western countries. India and chiefly Bengal, as mainly an agricultural country, requires its chief sources of income to be placed in the hands of the educated young men, and I quite agree with your views expressed in your evidence before the Agricultural Commission.

Source: 'Settlement of Khas Mahal land in Faridpur with young men of the bhadralok class', prog no 1-2, GoB, progs of rev dept (land rev), Jan 1928, in For the Quarter Ending March 1928.

Appendix 7: changes in the rainfall pattern in Bengal (inches)

Districts	Average between 1862–72	Average between 1911–21
Bogra	66.62	63.05
Midnapur	72.02	60.21
Hooghly	65.23	57.15
Murshidababd	65.62	55.37
Pabna	69.20	59.67
Jessore	72.15	62.37
Nadia	64.91	54.69
Burdwan	60.31	55.93
Tippera	93.50	82.01

Source: Abul Hussain, *The Problems of Rivers in Bengal* (Dacca, 1927), p. 24.

Appendix 8: statistical examples of changes in the land ownership, cultivators and non-cultivators (acres)

(a) Bakarganj

Thana	Areas held by non-cultivators (1918)	Areas held by non-cultivators (1933)
Lalmohan	5120	6400
Amtoli	1280	2560
Barisal	640	1280
Total	7040	10240

(b) Noakhali

Thana	Areas held by non-cultivators (1918)	Areas held by non-cultivators (1933)
Sandip	2042	2937
Raipur	692	1197
North Hatiya	580	1775
South Hatiya	1076	1680
Total	4390	7591

Source: Rai R.C. Sen Bahadur, Land Revenue Settlement Officer, Chittagong to Director of Land Records, Bengal, 2 August 1933, no. 1-8-2642. file no: 55C – 13/33(1), pp. 1–2, in NAB: Judicial Department, GoB, List: 114, Bundle: 2.

Bibliography

Manuscript sources

Asia, Pacific and Africa Collections, British Library, London (APAC)

European Manuscripts

Dash (A.) Memoirs.
Pinnel Papers.
Railway Consolidation Act, 1845.
Temple (Richard) Collection.

India Office Records

India and Bengal Despatches.
Proceedings (revenue) of the Government of Bengal.
Proceedings (land) of the Government of Bengal.
Proceedings (Judicial) of the Government of Bengal.
Report on the Administration of the Lower Provinces.
Trial of Doodoo Meea and his Followers, 1847 (Calcutta: 1848).

University Library, Cambridge (UL)

'Correspondences Relating to the Wahabi Movement'.
'Memorandum by F.D. Chauntrell, Solicitor to Government of India, on Government's defence against action for wrongful arrest, summarizing recent history of Wahabi Movement in India and previous trials of its members.'

Centre of South Asian Studies, Cambridge (CSAS)

Mukherjee (S.N.) Papers.
Bell (John) Papers.
Benthall Papers.
Donovan (J.T.) Papers.
Holland (E.W.) Papers.
Martin (O.M.) Papers.
Pinnell Papers.

Churchill College, Cambridge

Churchill Archives.

Faridpur and Comilla District Collectorate Record Rooms

Tauzi Records.

National Archives of Bangladesh, Dhaka (NAB)

Proceedings of the Communication, Buildings and Works Department.
Proceedings of the Co-operative Department.
Proceedings of the Irrigation Department.
Proceedings of the Land Revenue Department.
Proceedings of the Public Health Department.
Proceedings (Revenue) of the Lieutenant-Governor of Bengal.

West Bengal State Archives, Kolkata (WBSA)

Proceedings of Agriculture Department.
Proceedings of the Revenue Department.

Official publications

Official publications without author's name

Act IX of 1847: Alluvial Land, Bengal: Assessment of New Lands.
Act XIII of 1936: The Bengal Water Hyacinth Act, 1936 and the Rules Thereunder.
Annual Report on the Administration of Bengal Presidency.
Annual Report of the Department of Agriculture, Bengal.
Annual Report of the Royal Botanic Garden Calcutta.
Annual Report on the Working of Co-Operative Societies in the Presidency of Bengal.
Annual Sanitary Report of the Province of Eastern Bengal and Assam.
Bengal Government Estates Manual, 1932.
Bengal Legislative Assembly Proceedings.
Bengal Legislative Council Proceedings.
Bengal Waste Lands Manual 1919.
Bengal Waste Lands Manual 1936.
Bengal Waterways Act 1934.
Memorandum of the Hon'ble Mr Justice H.B.L Braund on Events From March 1943 to the End of 1943 in Relation to the Food Situation in Bengal (Calcutta: Govt of India).
Movements of the People and Land Reclamation Schemes (Calcutta, 1885).
Papers Relating to Culturable Wastelands at the Disposal of Government (Calcutta, 1860).
Principal Heads of the History and Statistics of the Dacca Division (Calcutta: E.M. Lewis, Central Company Limited, 1868).
Report on the Administration of Bengal, 1872–1879.
Report of the Bengal Paddy and Rice Enquiry Committee (vol. 1, Government of Bengal, Alipore: Bengal Government Press, 1939).
Report of the Bengal Provincial Banking Enquiry Committee, 1929–30, vol. II, Evidence, part I (Calcutta, 1930).
Report on the Census of India (Bengal), 1871–1941.
Report on the Condition of the Lower Classes of Population in Bengal (Calcutta: Bengal Secretariat Press, 1888).
Report on the Embankments of the Rivers of Bengal (Calcutta: W. Ridsdale, Bengal Military Orphan Press, 1846).

Report of the Foodgrains Procurement Committee (Calcutta: Government of Bengal, 1944).

Report of the Government of Bengal Unemployment Enquiry Committee, vol. II, Written and Oral Evidence (Calcutta: Bengal Secretariat Book Depot, 1925).

Report on the Hydraulic Condition of the Area Affected by the North Bengal Floods, in Proceedings of Government of Bengal, Irrigation Department, Irrigation Branch, For the Quarter Ending June 1927, vol. IV.

Report of the Indian Famine Commission (Calcutta: Office of Superintendent of the Government, India, 1898).

Report of the Indigo Commission Appointed Under Act XI. of 1860: With the Minutes of Evidence Taken Before Them and Appendix (Calcutta, 1860).

Report on the Land Revenue Administration of the Lower Provinces, 1874–5.

Report of the Land Revenue Commission, Bengal (vols I–VI, Board of Revenue, Alipore, 1940–41).

Report of the Water Hyacinth Committee 1922.

Resolution Reviewing the Reports on the Working of District and Local Boards in Bengal between 1932–33 and 1939–1940 (Calcutta: Bengal Government Press).

Settlement of Khasmahal Land With 'Bhadraloke' Youths in Faridpur, 1933 [Rajshahi, Rajshahi Commissioner's Library, shelfmark: 16(A)–42].

A Short Survey of the Work Achievements and Needs to the Bengal Agriculture Department, 1906–1936 (Calcutta: Government of Bengal, n.d).

Official publications with author's name

De, Kiran Chandra, *Report on the Fisheries of Eastern Bengal and Assam* (Shillong: Secretariat Printing Office, 1910).

Gastrell, G.E., *Geographical and Statistical Report of the Districts of Jessore, Fureedpore and Backergunge* (Calcutta: Superintendent of Government Printing Press, 1868).

Gupta, G.N., *A Survey of the Industries and Resources of Eastern Bengal and Assam for 1907–1908* (Shillong: Eastern Bengal and Assam Secretariat Printing Office, 1908).

Gupta, Pomode Ranjan Das, Settlement Officer of Faridpur, *Faridpur Revisional Settlement Final Report, Part-1 – Survey and Settlement Operations 1940–4* (Government of East Bengal, Dacca: East Bengal Government Press, 1954).

Harrison, Henry Leland, *The Bengal Embankment Manual Containing An Account of the Action of the Government in Dealing With Embankments And Water Courses Since the Permanent Settlement: Discussion of the Principles of Act of 1873* (Calcutta: The Bengal Secretariat Press, 1909).

Ishaque, H.S.M., *Agricultural Statistics by Plot to Plot Enumeration in Bengal 1944 and 1945*, Part II (Alipore: West Bengal Government Press, 1947).

Jack, J.C., *Final Report on the Survey and Settlement Operation in the Bakarganj District, 1900–1908* (Calcutta, 1915).

Kerr, Hem Chunder, *Report on the Cultivation of, and Trade in, Jute in Bengal*, (Calcutta: Bengal Secretariat Press, 1877).

Kindersley, J.B., *Final Report on the Survey and Settlement Operations in the District of Chittagong 1923–33* (Alipore: Bengal Government Press, 1938).

Mitchell, K.G. and L.H. Kirkness, *Report on the Present State of Road and Railways Competition and the Possibilities of Their Future Co-ordination and Development*, *and Cognate Matters, in Governors Provinces* (Calcutta: Government of India, Central Publications Branch, 1933).

Nelson, W.H., *Final Report on the Survey and Settlement Operation in the District of Rajshahi, 1912–1922* (Calcutta: Bengal Secretariat Book Depot, 1923).

O'Malley, L.S.S., *Bengal District Gazetteer: Pabna* (Calcutta: Bengal Secretariat Book Depot, 1923).

O'Malley, L.S.S., *Bengal District Gazetteers: Faridpur* (Calcutta: Bengal Secretariat Book Depot, 1925).

Prance, B.C., *Final Report on the Survey and Settlement Operations in the Riparian Area of District Tippera Conducted With the Faridpur District Settlement 1909 to 1915* (Calcutta: The Bengal Secretariat Book Depot, 1916).

Rickett, Henry, *Settlement Report [on Chittagong] of 1849*.

Stuart, M.M., *Khasmahal Report* (Alipore, 1938).

Webster, J.E., *Eastern Bengal District Gazetteers: Noakhali* (Allahabad: Pioneer Press, 1911).

Webster, J.E., *Eastern Bengal District Gazetteer: Tippera* (Allahabad: Pioneer Press, 1910).

Books and articles published before 1947

Bangla language sources

Aftab-i-hidayat (Mymensingh, 1877).

Ali, Muhammad Dad, *Samaj Siksha* [Instructions on Society] (Nadia: Muhammad Usaf Ali, 1910).

Ali, Siddik, *Siddik Alir Puthi*, 1244 BS (AD 1837).

Bandyopadhyaya, Bhabani Charan, *Kalikata Kamalalaya* (Calcutta: Ranjan Publishing House, 1938).

Basu, Shantipriya, *Banglar Chashi* [Bengal peasantry] (Calcutta: Biswa Bharati Granthalaya, 1944 [1351 BS]).

Bhatcharya, Biseshwar, 'Gramer Kotha' [Thoughts on villages], *Bangabani*, 4 (Baishakh 1332 BS).

Bhatacharya, Bisheswar, 'Bange Ganga' [The Ganges in Bengal], *Bangabani*, 3(2) (Ashwin, 1331 BS).

Bhatcharya, Ramhari, *Banchibar Upay* [Means to Live] (Jessore, 1332 BS [AD 1925]).

Bidyanidhi, Jogesh Chandra Roy, 'Annachinta' [Thoughts on Subsistence], *Prabashi* (Ashwin, 1332 BS), vol. 25, part. 1, no. 6.

Bose, Rai Chunilal, *Palli-swashtha O saral swashtha-bidhana* [Village Sanitation and a Manual of Hygiene], 4th edn (Calcutta: A.P. Basu, 1934).

Chakravarti, Rajkumar and Ananga Mohana Dasa, *Sandwiper Itihasa* [History of Sandweep] (Calcutta, 1922 [1330 BS]).

Chatterjee, Bankim Chandra, 'Bangalir Bahubal' in *Pravandha Pustaka* (Katalpara, Bangadarshan Jantraloy, 1880).

Gangopadhaya, Narayan, 'Upanibesh' [The Colony], in *Narayan Gangopadhaya Rachanabali*, Vol. 2, 2nd edn (Calcutta: Mitra and Ghose Publishers, 1386 BS).

Ghazi kalu and champaboti kanyar puthi (Mymensingh, 1870).

Hye, A.F.M. Abdul, *Adarsha Krishak* [The Ideal Cultivator], 2nd edn (Mymensingh, 1328 BS [AD 1921]).

Islam-Chitra O Samaj-Chitro (Sketches of Islam and the Muslim Society Dealing with the Present Degenerate Condition of the Muhammadans) (Mymensingh, 1914).

Majumdar, Bhava Ranjan (ed.), *Deser Gana* [Songs for the Country], 2nd edn (Barisal, 1905).

Majumdar, Jadunath, *Hindu Samajer Samashya* [Problems of the Hindu Society] (Jessore: Hindu Patrika, 1325 BS).

Mitra, Dina Bandhu, *Nila darpana nataka* [Indigo Mirror] (Calcutta, 1861).

Nag, Abinash Chandara, *Krishak o Sramajibi*, [Peasants and Labourers] (Calcutta: Sribasantakumar Mitra, 1907).

Putatunda, Brindabon Chadra, *Nuton Banger Puraton Kahini* [Old Stories of New Bengal] (Barisal,1323 BS).

Rahim, Abdur, *Akaler Puthi* [Puthi of Famine] (Mymensingh, 1875).

Rajanikanta, Pandit, *Swadeshi Palli Sangeet* [Swadeshi Country Songs] (Mymensingh, 1907).

Ray, Nikhila Nath, *Sonar Bangla* [Golden Bengal] (1906).

Sarkar, Pitambar, *Jati-Vikash* [Evolution of Caste] (A Treatise on the Caste System in Bengal, with Reference to the Caste Name Chasadhola) (Calcutta, 1910).

Sen, Navin Chandra, 'Amar Jivan', in Sajanikanta Das (ed.), *Navin Chandra Rachanalbali*, Vol. II, new edn (Calcutta, 1366 BS [AD 1959]).

Sen, Rishikesh, *Baker Samasya* [Unemployment Problem] (Chandannagor: Rameshwar and Co., 1934).

Sobhan, Sheikh Abdus, *Hindu-Mosolman* (Dacca: Azizon Library, 1889).

Sonar Vanla: Forty Collected Songs (Calcutta: Sanyal and Co, 1905).

Ullah, Mohammed Mohsen, *Burir Suta* (Rajshahi, Calcutta: 1317 BS).

English language sources

Addams-Williams, C., *Notes on the Lectures of Sir William Willcocks on the irrigation in Bengal together with a reply by Sir William Willcocks* (Calcutta: Bengal Secretariat Book Depot, 1931).

Alexander, James, 'On the Tenures and Fiscal Relations of the Owners, and Occupants of the Soil in Bengal, Behar, and Orissa', *Journal of the Asiatic Society of Bengal*, XIV (July–December 1845).

Allen, Isaac, 'The Revival of Islam', *The Calcutta Review* (January 1874).

'A Lover of Justice', *Permanent Settlement Imperilled or, Act X. of 1859 in Its True Colors* (Calcutta: The Englishman Press 1859).

An Old Indian Postmaster, *Indian Railways; As Connected with the Power and Stability of the British Empire in the East etc.* (London: Thomas Cautley Newby, 1846).

Ascoli, F.D., *A Revenue History of the Sundarbans From 1870–1920* (Calcutta: Bengal Secretariat Book Depot, 1921).

A Sketch of Eastern Bengal with Reference to Its Railways and Government Control (Calcutta: Thacker, Spink and Co., 1861).

Baden-Powell, B.H., *The Origin and Growth of Village Communities in India* (London: Swan Sonnenschein and Co., 1899).

Banerjee, Benoyendranath, 'Some Economic Problems of Bengal – 1: The Water Hyacinth', *The Bengal Co-operative Journal*, XIX(1) (1933).

Basu, K.P. and M.N. Basak., 'Biochemical Investigations on Different Varieties of Bengal Rice – Part V', *Indian Journal of Medical Research*, 24(4) (April 1937).

Basu, K.P. and H.N. De, 'Nutritional Investigation of Some Species of Bengal Fish', *Indian Journal of Medical Research*, 26(1) (July 1938).

Bentley, C.A., *Malaria and Agriculture in Bengal. How to Reduce Malaria in Bengal by Irrigation* (Calcutta: Bengal Secretariat Book Depot, 1925).

Beveridge, Henry, *The District of Bakarganj: Its History and Statistics* (London: Trübner & Co., 1876).

Bishop, T.H., 'The Working of the Cholera Prevention Scheme on the Lower Ganges Bridge Construction', *Indian Journal of Medical Research*, 1(2) (1913–14).

Bose, Chunilal, *Food* (Calcutta: Calcutta University, 1930).

Burrard, Sidney, 'Movements of the Ground Level in Bengal', *Royal Engineers Journal*, XLVII (1933).

Campbell, George, *Memoirs of My Indian Career*, ed. Charles E. Bernards, Vol. II (London: Macmillan and Co., 1893).

Carstairs, R., *The Little World of an Indian District Officer* (London: Macmillan and Co., 1912).

Chaudhuri, J., 'Agrarian Problems of Bengal – 1', *Bengal Co-operative Journal*, XXI, 13 (1936).

Christian, John L., 'Anglo-French Rivalry in Southeast Asia: its Historical Geography and Diplomatic Climate', *Geographical Review*, 31(2) (1941).

Clay, A.L., *Leaves From a Diary in Lower Bengal* (London, 1896).

Cort, R., *The Anti-Rail-Road Journal; or, Rail-Road Impositions Detected* (London: W. Lake, 1835).

Cotton, Arthur, 'On Communication between India and China by the Line of the Burhampooter', in the *Proceedings of the Royal Geographical Society of London*, 11(6) (1866–67).

Cotton, H.J.S., 'The Rice Trade in Bengal', *Calcutta Review*, CXV (January 1874).

Dey, Kanny Loll, *Hindu Social Laws and Habits Viewed in Relation to Health* (Calcutta: R.C. Lepage and Co, 1866).

Day, Lal Behari, *Bengal Peasant Life* (London, 1892).

Drucquer, Seith, 'On the Rivers of East Bengal', *Geographical Magazine*, XII(5) (March 1941).

Dutt, R.C., *The Peasantry of Bengal*, ed. Narahari Kaviraj (1874; reprinted Calcutta: Manisha Granthalaya, 1980).

Dutt, R.C., *The Economic History of India. In the Victorian Age*, Vol. II (London: Kegan Paul, Trench and Trubner, 1904).

Finney, S., *Railway Construction in Bengal. Three Lectures Delivered at the Sibpur Engineering College in January–February 1896* (Calcutta: Bengal Secretariat Press, 1896).

Fry, A.B., 'Indigenous Fish and Mosquito Larvae: a Note from Bengal', *Paludism: Being the Transaction of the Committee for the Study of Malaria in India*, 1(5) (September, 1912).

Ganguli, Birendranath, *Trends of Agriculture and Population in the Ganges Valley: a Study in Agricultural Economics* (London: Methuen & Co.,1938).

Geddes, Arthur, 'The Population of Bengal, its Distribution and Changes: a Contribution to Geographical Method', *Geographical Journal*, 89(4) (April, 1937).

Gandhi, M.K., *Collected Works of Mahatma Gandhi*, Vols 4–5 (Government of India: Ministry of Information and Broadcasting, 1988).

Grant, Colesworthey, *Rural Life in Bengal* (London: W. Thacker and Co., 1860).

Hollingbery, R.H., *The Zemindary Settlement of Bengal*, Vol. 1 (Calcutta: Brown & Co., 1879; reprinted Delhi, 1985).

Hooghly to the Himalayas, Being an Illustrated Handbook to the Chief Places of Interest Reached by the Eastern Bengal State Railway (Bombay: Times Press, 1913).

Hooker, Joseph Dalton, *Himalayan Journals* (London: John Murray, 1854).

Hunter, W.W., *The Indian Musalmans: are They Bound in Conscience to Rebel against the Queen?* (London: Trübner, 1871).

Hunter, W.W., *Famine Aspects of Bengal Districts* (London: Trübner, 1874).

Hunter, W.W., *A Statistical Account of Bengal*, Vols 1–6 (London: Trübner & Co., 1875).

Hunter, W.W. (ed.), *Imperial Gazetteer of India*, Vols I, IV, VI, IX (London: Trübner & Co., 1881–85).

Hunter, W.W., *The Indian Empire: its People, History, and Products* (London: Allen & Co., 1893).

Huque, M. Azizul, *The Man Behind the Plough* (Calcutta: Book Co., 1939).

Hussain, Abul, *The Problem of Rivers in Bengal* (Dacca, 1927).

Iyengar, M.O.T., 'The Distribution of *Anopheles Ludlowii* in Bengal and its Importance in Malarial Epidemiology', *Indian Journal of Medical Research*, 19(2) (October, 1931).

Jack, J.C., *Final Report on the Survey and Settlement Operation in the Bakarganj District, 1900–1908* (Calcutta: Bengal Secretariat Book Depot, 1915).

Jack, J.C., *The Economic Life of a Bengal District: a Study* (Oxford: Clarendon Press, 1916).

Joyce, H.W., *Five Lectures on Indian Railway Construction and one Lecture in Management Control* (Calcutta: The Bengal Secretariat Press, 1905).

Kennedy, J.P., *A Railway Caution!! Or Exposition of Changes Required in the Law & Practice of the British Empire etc.* (Calcutta: Messrs R.C. Lepage and Co., 1849).

Latif, S.A., *Economic Aspect of the Indian Rice Export Trade* (Calcutta: Dasgupta & Co., 1923).

Lees, O.C., *Waterways in Bengal: their Economic Value and the Methods Employed for their Improvement* (Calcutta: Bengal Secretariat Book Depot, 1906).

Mahalanobis, P.C., 'The Bengal Famine: the Background and Basic Facts', *Asiatic Review*, XLII (January 1946).

McLean, Kenneth, 'Water Hyacinth: a Serious Pest in Bengal', *Agricultural Journal of India*, XVII (1922).

M'Clelland, John, *Sketch of the Medical Topography or Climate and Soils, of Bengal and the N.W. Province* (London: John Churchill,1859).

Mill, James, *The History of British India*, Vol. 5 (London: Baldwin, Cradock, and Joy, 1826).

Mitra, B. B., *Laws of Land and Water in Bengal and Bihar* (Calcutta: Eastern Law House, 1934).

Mitter, S.C., *A Recovery Plan for Bengal* (Calcutta: The Book Company Limited, 1934).

Mookerjee, Radharomon, *History and Incidents of Occupancy Right* (Delhi: Neeraj Publishing House, 1919; reprinted 1984).

Mukeerji, Krishnakali, 'The Transferability of Occupancy Holding in Bengal, Part I', *Bengal Economic Journal*, 1(3) (January 1917).

Mukerjee, Radhakamal, *The Changing Face of Bengal: a Study in Riverine Economy* (Calcutta: Calcutta University, 1938).

Nivedita, Sister, *Glimpses of Famine and Flood in East Bengal in 1906* (Calcutta, 1907).

Niyogi, J.P., *The Co-Operative Movement in Bengal* (London : Macmillan and Co., 1940).

Panandikar, S.G., *The Wealth and Welfare of the Bengal Delta: Comprising the Districts of Mymensingh, Dacca, Bogra, Pabna, Faridpur, Bakarganj, Tippera and Noakhali* (Calcutta: Calcutta University Press, 1926).

Pargiter, F.E., *A Revenue History of the Sunderbans From 1765 to 1870* (Calcutta: Bengal Secretariat Press, 1885).

'Permanent Settlement of the Indian Land Revenue', in *The Asiatic Journal and Monthly Miscellany* (1929).

'Railways in India.—Bengal', *The Colonial Magazine & East India Review*, XVI (1849).

Rennell, James, 'Journal of Major James Rennell', ed. T.H.D. La Touche, in *Bengal Asiatic Society Memoirs*, III (1910–14).

Roy, B. (ed.), *Speeches and Addresses of Sir John Anderson* (London: Macmillan, 1939).

Roy, Baboo Isser Chunder Mitter, 'A Few Facts Concerning Village Life', The Bengal Social Science Association Meeting, 1877 (Calcutta: Wyman and Co., 1877).

Roy, M N, 'Bourgeois Nationalism', *Vanguard*, 3(1) (1923).

Roy, Raja Ram Mohun, *Exposition of the practical operation of the judicial and revenue systems of India: and of the general character and condition of its native inhabitants, as submitted in evidence to the authorities in England* (London: Smith, Elder & Co., 1832).

Roychowdhury, Birendra Kishore, *Permanent Settlment and After* (Calcutta: The Book Company Ltd., 1942).

Sadeque, A., 'The Co-operative Credit Movement and Interest Rate in India', *Bengal Co-Operative Journal*, XXV(2) (1939).

Sanyal, Hireankumar, 'Co-Operation in Bengal', *Bengal Co-Operative Journal*, XVIII(4) (1933).

Seton-Karr, W.S., 'Agriculture in Lower Bengal', *Journal of the Society of Arts*, 16 (March 1883).

Taylor, James, *A Sketch of the Topography and Statistics of Dacca* (Calcutta: Military Orphan Press, 1840).

Temple, Richard, *Men and Events of my Time in India* (London: Macmillan, 1882).

'Transit', *A Letter to the Shareholders of the East Indian Railway and to the Commercial Capitalists of England and India* (London: Smith, Elder & Co., 1848).

Twining, William, *Clinical Illustrations of the More Important Diseases of Bengal*, Vol. 1, 2nd edn (Calcutta: Parbury, Allen & Co, 1835).

Weed, Alfred C., 'Another Factor in Mosquito Control', *Ecology*, 5(1) (January 1924).

Westland, James, *A Report on the District of Jessore: its Antiquities, its History, and its Commerce*, 2nd edn (Calcutta: Bengal Secretariat Press, 1874).

Williams, C. Addams, *History of the Rivers in the Gangetic Delta 1750–1918* (Calcutta, 1919; reprinted Dacca: East Pakistan Inland Water Transport Authority, 1966).

Wise, James, *Notes on the Races, Castes, and Trades of Eastern Bengal* (London: Harrison and Sons, 1883).

Wise, James, 'The Muhammadans of Eastern Bengal', *Journal of the Asiatic Society of Bengal*, 1(3) (1894).

Wolfe, Henry W., *People's Banks: a Record of Social and Economic Success*, 3rd edn (London: P.S. King and Sons, 1910).

Wood, W.H. Arden, 'Rivers and Man in the Indus-Ganges Alluvial Plain', *Scottish Geographical Magazine*, 40(1) (1924).

Newspapers and periodicals (Bengali and English)

Amrita Bazar Patrika, Calcutta.
Bangabani, Calcutta.
Bangasri, Calcutta.
The Bangladesh Observer, Dhaka.
The Bengal Times, Dacca.
Dacca Prakash, Dacca.
The Times, London.
East Bengal and Assam Era, Dacca.
Modhostha, Dacca.

Books and articles published since 1947

Bangla language sources

Ahmad, Muhibuddin, *Aposhhin ek sangrami pir Dudu Mia* [Dudu Miyan: an Uncompromising Resistance Leader] (Dhaka: Shariatia Library, 1992).

Adnan, Shapan, *Bangladesher Krishi Prashno: Bhumi Sanskar O Khas Jamir Odhikar Protishthai Gonoandoloner Bhumika* [The Agricultural Question in Bangladesh: the Role of Mass Movement in Land Reforms and the Establishment of Rights on Khas Lands] (Dhaka: History Department, University of Dhaka, 2008).

Bhadra, Gautam, *Iman O Nishan: Unish shotoke bangaly krishak chaitanyer ek adhyay, c. 1800–1850* (Calcutta: Subarnarekha for the Centre for Studies in Social, 1994).

Maksud, Syed Abul (ed.), *Maolana Abdul Hamid Khan Bhasani* (Dhaka: Bangla Academy, 1994).

Sheikh, M Abu Hanif et al., *Chalan beel anchaler nadimalar shankyatattik bissleshon* [Statistical Analysis of the Rivers of the Chalan Beel], *IBS Journal*, 7 (1406 BS).

English language sources

Abdullah, Abu et al., 'Agrarian Structure and the IRDP – Preliminary Considerations', *Bangladesh Development Studies*, 4(2) (1976).

Adas, Michael, *The Burma Delta: Economic Development and Social Change on an Asian Rice Frontier, 1852–1941* (Madison: University of Wisconsin Press, 1974).

Adnan, Shapan and H. Zillur Rahman, 'Peasant Classes and Land Mobility: Structural Reproduction and Change in Rural Bangladesh', *Bangladesh Historical Studies*, 3 (1978).

Ahmad, Muzaffar, *Myself and the Communist Party of India*, trans. P.K. Sinha (Calcutta: National Book Agency, 1970).

Ahmed, Rafiuddin, *The Bengal Muslims 1871–1906: a Quest for Identity* (Delhi: Oxford University Press, 1981).

Ahmed, Sufia, *Muslim Community in Bengal, 1884–1912* (Dacca: S. Ahmed; distributed by Oxford University Press, Bangladesh, 1974).

Agrawal, Arun and K. Sivaramakrishnan, *Agrarian Environment. Resources, Representation, and Rule in India*, (Durham. NC: Duke University Press, 2000).

Alagh, Yoginder K. et al. (eds), *Interlinking of Rivers in India: Overview and Ken-Betwa Link* (New Delhi: Academic Foundation, 2006).

Alavi, Hamza, 'Peasants and Revolution', *Socialist Register* (1965).

Amrith, Sunil, *Decolonizing International Health: India and Southeast Asia 1930–65* (Basingstoke: Palgrave Macmillan, 2006).

Arnold, David, *Famine: Social Crisis and Historical Change* (Oxford: Basil Blackwell, 1988).

Arnold, David, *Colonizing the Body: State Medicine and Epidemic Disease in Nineteenth-Century India* (Berkeley: University of California Press, 1993).

Arnold, David, *Science, Technology and Medicine in Colonial India*, The New Cambridge History of India (Cambridge: Cambridge University Press, 2000).

Arnold, David and Ramachandra Guha (eds), *Nature, Culture, Imperialism: Essays on the Environmental History of South Asia* (Delhi: Oxford University Press, 1995).

Bandyopadhyay, J. and S. Perveen, 'Interlinking of Rivers in India: Assessing the Justifications', *Economic and Political Weekly*, 39(50) (2004).

Bandyopadhyay, J. and S. Perveen, 'Doubts over the Scientific Validity of the Justifications for the Proposed Interlinking of Rivers in India', *Science and Culture*, 70(1–2) (2004).

Bandyopadhyay, Sekhar, *Caste, Protest and Identity in Colonial India: the Namasudras of Bengal, 1872–1947* (London: Curzon, 1997).

Bandyopadhyay, Sekhar (ed.), *Bengal: Rethinking History. Essays in Historiography* (New Delhi: Manohar, 2001).

Barkat, Abul et al. (eds), *Political Economy of Khas Land in Bangladesh* (Dhaka: Association for Land Reform and Development, 2001).

Barton, Gregory, *Empire Forestry and the Origins of Environmentalism* (Cambridge: Cambridge University Press, 2002).

Bayly, C.A., *Rulers, Townsmen and Bazaars: North Indian Society in the Age of British Expansion 1770–1870* (Cambridge: Cambridge University Press, 1983).

Bayly, C.A., 'State and Economy in India over Seven Hundred Years', *Economic History Review*, n.s., 38(4) (1985).

Bayly, C.A., 'Creating a Colonial Peasantry: India and Java c. 1820–1880', *Itinerario*, XI(I) (1987).

Bayly, C.A., *The Origins of Nationality in South Asia. Patriotism and Ethical Government in the Making of Modern India* (Delhi: Oxford University Press, 1998).

Bayly, Susan, *Caste, Society and Politics in India from the Eighteenth Century to the Modern Age* (Cambridge: Cambridge University Press, 1999).

Bear, Laura, *Lines of the Nation: Indian Railway Workers, Bureaucracy, and the Intimate Historical Self* (New York: Columbia University Press, 2007).

Beinart, William and Lotte Hughes, *Environment and Empire* (Oxford: Oxford University Press, 2007).

Bertocci, Peter, 'Structural Fragmentation and Peasant Classes in Bangladesh', *Journal of Social Studies*, 5 (1979).

Bhatia, B.M., *Famine in India: a Study of some Aspects of the Economic History of India 1860–1965* (Bombay, 1963; 2nd edn, 1967).

Bhattacharya, Tithi, *The Sentinels of Culture: Class, Education, and the Colonial Intellectual in Bengal (1848–85)* (New Delhi, Oxford University Press, 2005).

Blair, Harry W., 'Local Government and Rural Development in the Bengal Sundarbans: an Enquiry in Managing Common Property Resources', *Agriculture and Human Values*, 7(2) (1990).

Blyn, George, *Agricultural Trend in India* (Philadelphia, 1966).

Bose, Sugata, 'The Roots of "Communal" Violence in Rural Bengal: a Study of the Kishoreganj Riots, 1930', *Modern Asian Studies*, 16(3) (1982).

Bose, Sugata, *Agrarian Bengal: Economy, Social Structure and Politics, 1919–1947* (Cambridge: Cambridge University Press, 1986).

Bose, Sugata, *Peasant Labour and Colonial Capital: Rural Bengal since 1770*, The New Cambridge History of India (Cambridge: Cambridge University Press, 1993).

Boyce, James K., *Agrarian Impasse in Bengal: Institutional Constraints to Technological Change* (Oxford: Oxford University Press, 1987).

Broomfield, J.H., *Elite Conflict in a Plural Society: Twentieth-Century Bengal* (Berkeley: University of California Press, 1968).

Catanach, Ian J., *Rural Credit in Western India 1875–1930: Rural Credit and Co-operative Movement in the Bombay Presidency* (Berkeley: University of California Press, 1970).

Cederlöf, Gunnel, *Landscapes and the Law: Environmental Politics, Regional Histories, and Contests over Nature* (Ranikhet, New Delhi: Permanent Black, 2008).

Cederlöf, Gunnel and K. Sivaramakrishnan (eds), *Ecological Nationalisms: Nature, Livelihoods, and Identities in South Asia* (Seattle: University of Washington Press, 2006).

Chakrabarty, Bidyut, *The Partition of Bengal and Assam, 1932–1947: Contour of Freedom* (London and New York: Routledge/Curzon, 2004).

Chakraborty, Ratan Lal and Haruo Noma (eds), 'Selected Records on Agriculture and Economy of Comilla District, 1782–1867', JSARD Working Paper no. 13, Dhaka, 1989.

Chatterji, Joya, *Bengal Divided: Hindu Communalism and Partition, 1932–1947* (Cambridge: Cambridge University Press, 1994).

Chatterji, Joya, 'The Fashioning of a Frontier: the Radcliffe Line and Bengal's Border Landscape 1947–52', *Modern Asian Studies*, 33(1) (1999).

Chatterjee, Partha, 'The Colonial State and Peasant Resistance in Bengal 1920–1947', *Past and Present*, 110 (1986).

Chatterjee, Partha, *The Present History of West Bengal: Essays in Political Criticism* (Delhi: Oxford University Press, 1997).

Chatterjee, S.P., *Bengal in Maps* (Calcutta, 1949).

Chaudhuri, Binay Bhushan, 'Growth of Commercial Agriculture in Bengal – 1859–1885', *Indian Economic and Social History Review*, 7(1) (1970).

Chaudhuri, Binay, 'Agricultural Production in Bengal, 1850–1900: Co-existence of Decline and Growth', *Bengal Past and Present*, 88(2) (1969).

Chaudhuri, Binay, 'The Process of Depeasantization in Bengal and Bihar, 1885–1947', *Indian Historical Review*, 21(1) (1975).

Choudhury, Nurul H., *Peasant Radicalism in Nineteenth Century Bengal: the Faraizi, Indigo and Pabna movements* (Dhaka: Asiatic Society of Bangladesh, 2001).

Choudhury, Nusha Yamina, Alak Paul and Bimal Kanti Paul, 'Impact of Coastal Embankment on the Flash Flood in Bangladesh', *Applied Geography*, 24(3) (July 2004).

Chaudhury, Sushil, *From Prosperity to Decline: Eighteenth Century Bengal* (New Delhi: Manohar, 1995).

Collins, William J., 'Labor Mobility, Market Integration, and Wage Convergence in Late 19th Century India', *Explorations in Economic History*, 36(3) (July 1999).

Cooper, Adrienne, *Sharecropping and Sharecroppers' Struggles in Bengal 1930–1950* (Calcutta: K.P. Bagchi & Company, 1988).

Crosby, Alfred W., *The Colombian Exchange: Biological and Cultural Consequences of 1492* (Westport, CT: Greenwood, 1972).

Crosby, Alfred W., *Ecological Imperialism: the Biological Expansion of Europe, 900–1900* (Cambridge: Cambridge University Press, 1986).

D'Souza, Rohan, *Drowned and Dammed: Colonial Capitalism, and Flood Control in Eastern India* (Delhi: Oxford University Press, 2006).

Damodraran, Vinita, 'Famine in Bengal: a Comparison of the 1770 Famine in Bengal and the 1897 Famine in Chotanagpur', *Medieval History Journal*, 10(1–2) (2007).

Datta, Rajat, *Society, Economy and Market: Commercialization in Rural Bengal c.1760–1800* (Delhi: Manohar Publishers, 2000).

Davis, Mike, *Late Victorian Holocausts: El Niño, Famines and the Making of the Third World* (London and New York: Verso, 2001).

De, Amalendu, 'Sri Aurobindo's Role in Indian Freedom Struggle: an Assessment from Different Perceptions', Presidential Address 2002–03 (Kolkata: The Asiatic Society).

Derbyshire, I.D., 'Economic Change and the Railways in North India, 1860–1914', *Modern Asian Studies*, 21(3) (1987).

Desai, A.R. (ed.), *Peasant Struggles in India* (Bombay: Oxford University Press, 1979).

Devereux, Stephen, 'Sen's Entitlement Approach. Critique and Counter-Critiques', *Oxford Development Studies*, 29(2) (2001).

Drayton, Richard, *Nature's Government: Science, Imperial Britain, and the 'Improvement' of the World* (New Haven and London: Yale University Press, 2000).

Dyson, Tim, 'On the Demography of South Asian Famines, part II', *Population Studies*, 45(2) (July 1991).

Dyson, Tim and Arup Maharatna, 'Excess Mortality during the Bengal Famine: a Re-Evaluation', *Indian Economic and Social History Review*, 28 (1991).

Eaton, Richard M., *The Rise of Islam and the Bengal Frontier 1204–1760* (Berkeley, Los Angeles and London: University of California Press, 1993).

Eaton, Richard M., 'Human Settlement and Colonization in the Sundarbans', *Agriculture and Human Values*, 7(2) (1990).

Edney, Matthew H., *Mapping an Empire: the Geographical Construction of British India, 1765–1843* (Chicago: University of Chicago Press, 1997).

Encyclopaedia of Islam, Vol. II (Leiden, London, 1965).

Feroze, Towheed, 'Land Policy: Pro-Poor Plan is the Key', *Dhaka Courier*, 6–12 June 2008.

Fieldhouse, D.K., 'Colonialism: Economic', *International Encyclopedia of Social Sciences* (New York: Macmillan, 1968).

Fischer-Tine, Harald and Michael Mann (eds), *Colonialism as Civilizing Mission: Cultural Ideology in British India* (London: Anthem Press, 2004).

Fitch, K.S., *A Medical History of the Bengal Famine 1943–44* (Calcutta: Government of India Press, 1947).

Gadgil, Madhav and Ramachandra Guha, *This Fissured Land: an Ecological History of India* (Delhi : Oxford University Press, 1992).

Gandhi, M.K. *Collected Works of Mahatma Gandhi*, Vol. 5 (Delhi: Government of India, 1984).

Geddes, Arthur, 'Alluvial Morphology of the Indo-Gangetic Plain', *Transactions and Papers* (Institute of British Geographers), 28 (1960).

Geertz, Clifford, *Agricultural Involution: the Processes of Ecological Change in Indonesia* (California: University of California Press, 1970).

Gidwani, V.K., '"Waste" and the Permanent Settlement in Bengal', *Economic and Political Weekly*, 27(4) (1992).

Ghosh, Devleena, Heather Goodall and Stephanie Hemelryk Donald (eds), *Water, Sovereignty, and Borders in Asia and Oceania* (London: Routledge, 2009).

Gopal, Sarvepalli, *The Permanent Settlement in Bengal and its Results* (London: G. Allen & Unwin, 1949).

Gramsci, Antonio, *Selections From the Prison Notebooks*, ed. and trans. Q. Hoare and G.N. Smith (London: Lawrence and Wishart, 1971).

Greenough, Paul R., *Prosperity and Misery in Modern Bengal* (New York and Oxford: Oxford University Press, 1982).

Grove, Richard, *Ecology, Climate and Empire: the Indian Legacy in Global Environmental History 1400–1940* (Delhi: Oxford University Press, 1998).

Grove, Richard, *Green Imperialism: Colonial Expansion, Tropical Island Edens, and the Origins of Environmentalism, 1600–1860* (Cambridge: Cambridge University Press, 1995).

Grove, Richard et al. (eds), *Nature and the Orient: the Environmental History of South and Southeast Asia* (Delhi: Oxford University Press, 1998).

Guha, Ramachandra (ed.), *Social Ecology* (Delhi: Oxford University Press, 1994).

Guha, Ramachandra, *The Unquiet Woods: Ecological Change and Peasant Resistance in the Himalaya* (Berkeley: University of California Press, 2000).

Guha, Ranajit, *Elementary Aspects of Peasant Insurgency in Colonial India* (Durham: Duke University Press, 1999).

Guha, Ranajit, *A Rule of Property for Bengal: an Essay on the Idea of Permanent Settlement* (Paris: Mouton, 1963).

Guha, Ranajit (ed.), *Subaltern Studies I: Writing on South Asian History and Society* (New Delhi: Oxford University Press, 1982).

Guha, Ranajit, *Dominance without Hegemony: History and Power in Colonial India* (Cambridge, MA, and London: Harvard University Press, 1997).

Guha, Sumit, 'Agrarian Bengal, 1850–1947: Issues and Problems', *Studies in History*, n.s., 11(1) (1995).

Hardiman, David (ed.), *Peasant Resistance in India 1858–1914* (Delhi and Oxford: Oxford University Press, 1992).

Hardiman, David, 'Power in the Forest: the Dang 1820–1920', in *Subaltern Studies VIII* (New Delhi, 1994).

Hasan, Mushirul and M. Asaduddin (eds), *Image and Representation: Stories of Muslim Lives in India* (Delhi: Oxford University Press, 2000).

Hashmi, Taj ul-Islam, *Pakistan as a Peasant Utopia: the Communalization of Class Politics in East Bengal, 1920–1947* (Boulder: Westview, 1992).

Hill, Christopher V., *South Asia: an Environmental History* (Santa Barbara, CA: ABC-CLIO, 2008).

Iftikhar-ul-Awal, A.Z.M., 'The Problem of Middle Class Educated Unemployment in Bengal, 1920–1942', *Indian Economic and Social History Review*, XIX(1) (1982).

Iqbal, Iftekhar, 'The Railway in Colonial India: between Ideas and Impacts', in Roopa Srinivasan et al. (eds), *Our Indian Railways* (New Delhi: Foundation Books, 2005).

Iqbal, Iftekhar, 'Modernity and its Discontents: Studying Environmental History of Colonial and Postcolonial Bangladesh', *East West Journal*, 1(1) (2007), Dhaka.

Islam, M. Mufakharul, *Bengal Agriculture 1920–1946* (Cambridge: Cambridge University Press, 1978).

Islam, M. Mufakharul, 'The Great Bengal Famine and the Question of FAD yet Again', *Modern Asian Studies*, 41(2) (2007).

Islam, Sirajul, 'Bengal Peasantry in Debt', *Dacca University Studies*, 22 (Pt. A) (1974).

Islam, Sirajul, *Permanent Settlement in Bengal 1790–1819* (Dhaka: Bangla Academy, 1979).

Islam, Sirajul, *Bengal Land Tenure: the Origin and Growth of Intermediate Interests in the 19th Century* (Calcutta: K.P. Bagchi & Co., 1988).

Islam, Sirajul, *Rent and Raiyat: Society and Economy of Eastern Bengal, 1859–1928* (Dhaka: Asiatic Society of Bangladesh, 1989).

Islam, Sirajul (ed.), *History of Bangladesh 1704–1971*, 3 vols (Dhaka: Asiatic Society of Bangladesh, 1997).

Islam, Sirajul (ed.), *Banglapedia* (Dhaka: Asiatic Society of Bangladesh, 2003).

Jalais, Anu, 'The Sundarbans: Whose World Heritage Site?', *Conservation and Society*, 5(3) (2007).

Joardar, Safiuddin, 'Tenancy and Land Revenue in the Non-Permanently Settled Areas of Rajshahi: Some Case Studies', *Journal of the Institute of Bangladesh Studies*, III (1978).

Johnson, Gordon, 'Partition Agitation and Congress: Bengal 1904–1908', *Modern Asian Studies*, 7(3) (1973).

Jones, Kenneth W., *Socio-Religious Reform Movements in British India*, The New Cambridge History of India (Cambridge: Cambridge University Press, 1989).

Kaida, Yoshihiro (ed.), *The Imagescape of Six Great Asian Deltas in the 21st Century* (Kyoto: Centre for Southeast Asian Studies, 2000).

Kamal, Ahmed, 'Living with Water: Bangladesh since Ancient Times', in T. Tvedt and E. Jakobsson (eds), *A History of Water*, Vol. 3 (London: I.B. Tauris Publishers, 2006).

Kamtekar, Indivar, 'A Different War Dance: State and Class in India 1939–1945', *Past and Present*, 176(1) (2002).

Kaviraj, Narahari, *Wahabi and Farazi Rebels of Bengal* (New Delhi: People's Publishing House, 1982).

Kawai, Akinobu, *Landlords and Imperial Rule: Change in Agrarian Bengal Society, c.1885–1940*, 2 vols (Tokyo: Institute for the Study of Languages and Cultures of Asia and Africa, 1986–87).

Kerr, Ian J. (ed.), *Themes in Indian History: Railways in Modern India* (Oxford: Oxford University Press, 2001).

Khalid, A.R.M., 'The Interlinking of Rivers Project in India and International Water Law: an Overview', *Chinese Journal of International Law*, 3(2) (2004).

Khan, Muin-ud-Din Ahmad, *History of the Fara'idi Movement in Bengal, 1818–1906* (Karachi: Pakistan Historical Society, 1965).

Khan, Muin-ud-din Ahmad, *Titu Mir and His Followers in British Indian Records* (Dacca: Barnamichhil, 1977).

Khan, Mushtaq Husain, 'Power, Property Rights and the Issue of Land Reform: a General Case Illustrated with Reference to Bangladesh', *Journal of Agrarian Change*, 4(1–2) (2004).

Klein, Ira, 'Death in India, 1871–1921', *Journal of Asian Studies*, 32(4) (August 1973).

Kling, Blair B., *The Blue Mutiny: Indigo Disturbances in Bengal, 1859–1862* (Philadelphia: University of Pennsylvania Press,1966).

Kopf, David (ed.), *Bengal* (Michigan: Michigan University Press, 1969).

Kumar, Dharma (ed.), *The Cambridge Economic History of India*, Vol. II (Cambridge, 1982).

Lapidus, Ira M., *A History of Islamic Societies* (Cambridge: Cambridge University Press, 1988).

Ludden, David (ed.), *Agricultural Production and Indian History* (Oxford: Oxford University Press, 1994).

Ludden, David, *An Agrarian History of South Asia* (Cambridge: Cambridge University Press, 1999).

Ludden, David (ed.), *Reading Subaltern Studies: Critical History, Contested Meaning, and the Globalization of South Asia* (Delhi: Permanent Black; London: Anthem Press, 2002).

Ludden, David (ed.), *Agricultural Production, South Asian History, and Development Studies* (New Delhi: Oxford University Press, 2005).

Maharatna, Arup, 'Malaria Ecology, Relief Provision and Regional Variation Mortality during the Bengal Famine of 1943–44', *South Asia Research*, 13(1) (May 1993).

Malik, M.B.K., *Hundred Years of Pakistan Railways* (Karachi: Government of Pakistan, Ministry of Railways and Communications, 1962).

Mallick, A.R., *British Policy and the Muslims in Bengal 1757–1856* (Dacca: Bangla Academy, 1977).

Mann, Michael, 'Mapping the Country: European Geography and the Cartographical Construction of India, 1760–90', *Science Technology & Society*, 8 (Mar 2003).

Marshall, P.J., *Bengal: the British Bridgehead*, The New Cambridge History of India (Cambridge: Cambridge University Press, 1988; reprinted 1990).

Mayaram, Shail, M.S.S. Pandian and Ajay Skaria (eds), *Subaltern Studies XII: Muslims, Dalits and the Fabrications of History* (New Delhi: Permanent Black and Ravi Dayal, 2005).

Misra, A.K. Saxena, A. Yaduvanshi, M. Mishra et al., 'Proposed River-Linking Project in India: Boon or Bane to Nature', *Environmental Geology*, 51(8) (2007).

Molla, M.K.U, *The New Province of Eastern Bengal & Assam, 1905–1911* (Rajshahi: Institute of Bangladesh Studies, Rajshahi University, 1981).

Mosse, David, *The Rule of Water: Statecraft, Ecology and Collective Action in South Asia* (Oxford: Oxford University Press, 2003).

Mukherjee, Hena, *The Early History of the East Indian Railway 1845–1879* (Calcutta: Firma KLM, 1994).

Mukherjhee, Mukul, 'Railways and their Impacts on Bengal's Economy: 1870–1920', *Indian Economic and Social History Review*, 17(2) (April–June 1980).

Mukherji, Saugata, 'Agrarian Class Formation in Modern Bengal 1931–1951', Occasional Paper no. 75, Calcutta: Centre for Studies in Social Sciences, 1985.

Neinary, William and Karen Middleton, 'Plant Transfers in Historical Perspective: a Review Article', *Environment and History*, 10 (2004).

Nakazato, Narikari, *Agrarian System in Eastern Bengal, c. 1870–1910* (Calcutta: K.P. Bagchi & Co.,1994).

Nandy, Srischandra, *Bengal Rivers and our Economic Welfare* (Calcutta: The Book Company Limited, 1948).

Nizami, Tauriq Ahmad, *Muslim Political Thought and Activity in India during the First Half of the 19th Century* (Aligarh: Three Mens Publications, 1969).

Nusha Yamina Choudhury, Alak Paul and Bimal Kanti Paul, 'Impact of Coastal Embankment on the Flash Flood in Bangladesh', *Applied Geography*, 24(3) (July 2004).

Odum, Eugene P., *Ecology and our Endangered Life-Support Systems* (Sunderland, MA: Sinauer Associates, 1989).

Osmani, S.R., 'The Entitlement Approach to Famine: an Assessment', Working Paper no. 107, Helsinki: UNU World Institute for Development Economics Research, 1993.

Padmanabhan, S.Y., 'The Great Bengal Famine', *Annual Review of Phytopathology*, 11 (1973).

Patkar, Medha (ed.), *A Millennium Folly?* (Mumbai: National Alliance for People's Movement and Initiatives, 2004).

Patnaik, Utsa, 'Food Availability and Famine: a Longer View', *Journal of Peasant Studies*, 19(1) (1991).

Popkins, Samuel L., *The Rational Peasant: the Political Economy of Rural Society in Vietnam* (Berkeley: University of California Press, 1979).

Prakash, Gyan, 'Subaltern Studies as Postcolonial Criticism', *American Historical Review*, 99 (December 1994).

Rajan, Ravi, *Modernizing Nature: Forestry and Imperial Eco-Development 1800–1950* (Oxford: Oxford University Press, 1996).

Rangarajan, Mahesh, *Fencing the Forest: Modernizing Nature. Forestry and Imperial Eco-Development 1800–1950* (New Delhi, Oxford and New York: Oxford University Press, 1999).

Rashid, Haroun er, *Geography of Bangladesh* (Dhaka: University Press, 1991).

Ray, Kabita, *History of Public Health: Colonial Bengal 1921–1947* (Calcutta: K.P. Bagchi and Company, 1998).

Ray, Rajat, *Social Conflict and Political Unrest in Bengal 1875–1927* (New Delhi: Oxford: Oxford University Press 1984).

Ray, Rajat and Ratnalekha Ray, 'Zamindars and Jotedars: a Study of Rural Politics in Bengal', *Modern Asian Studies*, 9(1) (1975).

Ray, Ratnalekha, *Change in Bengal Agrarian Society c. 1760–1850* (New Delhi: Manohar, 1979).

Richards, John F. and Elizabeth P. Flint, 'Long-Term Transformations in the Sundarbans Wetland Forests of Bengal', *Agriculture and Human Values*, 7(2) (1990).

Robb, Peter, 'Law and Agrarian Sciety in India: the Case of Bihar and the Nineteenth-Century Tenancy Debates', *Modern Asian Studies*, 22(2) (1988).

Robbins, Paul, *Political Ecology: a Critical Introduction* (Malden, MA: Blackwell, 2004).

Rogaly, Ben, Barbara Harriss-White and Sugata Bose (eds), *Sonar Bangla? Agricultural Growth and Agrarian Change in West Bengal and Bangladesh* (New Delhi: Sage, 1999).

Roy, Anuradha, *Nationalism as a Poetic Discourse in Nineteenth Century Bengal* (Calcutta: Papyrus, 2003).

Roy, Asim, *The Islamic Syncretistic Tradition in Bengal* (Princeton: Princeton University Press, 1983).

Roy, Tirthankar, 'Globalization, Factor Prices and Poverty in Colonial India', *Australian Economic Review*, 47(1) (2007).

Ruud, Arild Engelsen, 'The Indian Hierarchy: Culture, Ideology and Consciousness in Bengali Village Politics', *Modern Asian Studies*, 33(3) (1999).

Samanta, Arabinda, *Malarial Fever in Colonial Bengal 1820–1939: Social History of an Epidemic* (Kolkata: Firma KLM, 2002).

Schendel, Willem van, *Three Deltas: Accumulation and Poverty in Rural Burma, Bengal and South India* (New Delhi: Sage, 1991).

Schendel, Willem van, 'The Invention of the Jummas: State Formation and Ethnicity in Southeastern Bengal', *Modern Asian Studies*, 26(1) (1992).

Schendel, Willem van and Aminul Haque Faraizi, 'Rural Labourers in Bengal, 1880 to 1980', Comparative Asian Studies Program (CASP), Erasmus University Rotterdam, 1984.

Scott, James C., *Weapons of the Weak: Everyday Forms of Peasant Resistance* (New Haven and London: Yale University Press, 1985).

Scott, James C., *Seeing Like a State: How Certain Schemes to Improve the Human Condition Have Failed* (New Haven and London: Yale University Press, 1998).

Scott, James C. and Nina Bhatt (eds), *Agrarian Studies: Synthetic Work at the Cutting Edge* (New Haven: Yale University Press, 2001).

Seal, Anil, *The Emergence of Indian Nationalism: Competition and Collaboration in the Later Nineteenth Century* (Cambridge: Cambridge University Press, 1978).

Seidensticker, J. and A. Hai, *The Sundarbans Wildlife Management Plan: Conservation in the Bangladesh Coastal Zone* (Gland: IUCN, 1983).

Sen, Amartya Kumar, *Poverty and Famines: an Essay on Entitlement and Deprivation* (Oxford: Clarendon Press, 1981; reprinted 1988).

Sengupta, S.M., 'Interlinking of Rivers May Cause Geomorphic Changes', *Journal of Geological Society of India*, 69(5) (2007).

Sessions, George (ed.), *Deep Ecology for the 21st Century* (Boston: Shambala, 1995).

Shail Mayaram, M.S.S. Pandian and Ajay Skaria (eds), *Subaltern Studies XII: Muslims, Dalist and the Fabrications of History* (New Delhi: Permanent Black and Ravi Dayal, 2005).

Sharma, Sunil Sen (ed.), *Farakka – a Gordian Knot: Problems on Sharing Ganga Waters* (Calcutta: ISHIKA, 1986).

Singh, Chetan, *Natural Premises: Ecology and Peasant Life in the Western Himalaya, 1800–1950* (Delhi: Oxford University Press, 1998).

Singh, S.R. and M.P. Srivastava (eds), *River Interlinking in India: the Dream and Reality* (New Delhi: Deep and Deep, 2006).

Sivaramakrishnan, K., 'A Limited Forest Conservancy in Southwest Bengal, 1864–1912', *Journal of Asian Studies*, 56(1)(1997).

Sivaramkrishnan, K., *Modern Forests: Statemaking and Environmental Change in Colonial Eastern India* (Oxford: Oxford University Press, 1999).

Siok-Hwa, Cheng, *The Rice Industry of Burma 1852–1940* (Kuala Lumpur: University of Malaya Press, 1968).

Spira, W.M. et al., 'Uptake of *V. cholerae* Biotype El Tor from Contaminated Water by Water Hyacinth (*Eichornia crassipes*)', *Applied Environmental Microbiology*, 42 (1981).

Stein, Burton (ed.), *The Making of Agrarian Policy in British India 1770–1900* (Delhi: Oxford University Press, 1992).

Stewart, Gordon T., *Jute and Empire: the Calcutta Jute Wallahs and the Landscapes of Empire* (Manchester: Manchester University Press, 1998).

Stokes, Eric, 'The Return of the Peasant to South Asian History', *South Asia*, 6, 1 (1976).

Stokes, Eric, *The Peasant and the Raj* (Cambridge: Cambridge University Press, 1978).

Stone, Ian, *Canal Irrigation in British India: Perspectives on Technological Change in a Peasant Society* (Cambridge: Cambridge University Press, 1984).

Tauger, Mark B., 'Entitlement, Shortage and the 1943 Bengal Famine: Another Look', *Journal of Peasant Studies*, 31(1) (October 2003).

Thomasson, Jannuzi, F., *The Agrarian Structure of Bangladesh: an Impediment to Development* (Boulder, CO: Westview Press, 1980).

Thorner, Daniel, 'Capital Movement and Transportation: Great Britain and the Development of India's Railways', *Journal of Economic History*, 11(4) (1951).

Thorner, Daniel and Alice Thorner, *Land and Labour in India* (Bombay: Asia Publishing House, 1965).

Tomlinson, B.R., *The Economy of Modern India 1860–1970* (Cambridge: Cambridge University Press, 1996).

Tomlinson, B.R., 'Bengal Textiles, British Industrialisation, and the Company Raj: Muslins, Mules and Remittances, 1770–1820', *Bulletin of Asia-Pacific Studies No. X*, Kansai Institute of Asia-Pacific Studies, 31 March 2000.

Tucker, R.P. and J.F. Richards (eds), *Global Deforestation and the Nineteenth-Century World Economy* (Durham, NC: Duke University Press, 1983).

Tucker, Richard, 'Dimensions of Deforestation in Himalaya: the Historical Setting', *Mountain Research and Development*, VII(3) (1987).

Turner, B.L. II (ed.), *The Earth as Transformed by Human Action: Global and Regional Changes in the Biosphere over the Past 300 Years* (Cambridge: Cambridge University Press, 1990).

Uddin, Ashraf and Neil Lundberg, 'Cenozoic History of the Himalayan-Bengal System: Sand Composition in the Bengal Basin, Bangladesh', *Geological Society of America Bulletin*, 110(4) (April 1998).

Umar, Badruddin, *The Bengal Peasantry under the Permanent Settlement* (Dacca: Mawla Brothers, 1972) (in Bangla).

Veer, Peter van der and Hartmut Lehmann (eds), *Nation and Religion: Perspectives on Europe and Asia* (Princeton: Princeton University Press, 1999).

Viles, H. and T. Spencer, *Coastal Problems: Geomorphology, Ecology and Society at the Coast* (London: Edward Arnold, 1995).

Wallerstein, Immanuel, *The Modern World-System* (New York: Academic Press, 1974).

Wallerstein, Immanuel (ed.), *The Modern World-System in the Longue Durée* (Boulder and London: Paradigm, 2004).

Williams, Michael, *Deforesting the Earth: From Prehistory to Global Crisis, An Abridgment* (Chicago: Chicago University Press, 2006).

Wilson, Jon E., '"A Thousand Countries to go to": Peasants and Rulers in Late Eighteenth-Century Bengal', *Past & Present*, 189 (2005).

Wilson, Jon E., *The Domination of Strangers. Modern Governance in Eastern India* (Basingstoke: Palgrave, 2008).

Wolf, Eric, *Peasants* (New Jersey: Englewood Cliffs, 1966).

Woo-Cumings, Meredith, 'The Political Ecology of Famine: the North Korean Catastrophe and its Lessons', ADB Institute Research Paper 31, Tokyo: ADB Institute, 2002.

Woodroffe, Colin D., *Coasts: Form, Process and Evolution* (Cambridge: Cambridge University Press, 2002).

Unpublished doctoral dissertations

Bari, Muhammad Abdul, 'A Comparative Study of the Early Wahabi Doctrines and Contemporary Reform Movements in Indian Islam' (unpublished D.Phil. dissertation, University of Oxford, 1953).

Ihtesham, Kazi, 'Malaria in Bengal from 1860 to 1920: a Historical Study in a Colonial Setting' (unpublished PhD thesis, University of Michigan, 1986).

Lourdusamy, J., 'Science and National Consciousness: a Study of the Response to Modern Science in Colonial Bengal, c. 1870–1930' (unpublished D.Phil. thesis, University of Oxford, 1999).

Miki, Sayako, 'Salt Trade in Bengal in the Late 18th and Early 19th Centuries' (unpublished PhD thesis, SOAS, University of London, 2004).

Mukherjee, Tilottama, 'Markets, Transport and the State in the Bengal Economy, c. 1750–1800' (unpublished PhD thesis, University of Cambridge, 2004).

Samad, Abdus, 'Dynamics of Ascriptive Politics: a Study of Muslim Politicization in East Bengal (Bangladesh)' (unpublished PhD dissertation, Columbia University, 1983).

Non-governmental reports

Millennium Ecosystem Assessment, *Ecosystems and Human Well-Being: Synthesis* (Washington, DC: Island Press, 2005).

Internet sources

'Bangaldesh "is growing" due to freak environmental conditions', *Timesonline*, 1 August 2008, http://www.timesonline.co.uk/tol/news/environment/article 4440982.ece, last accessed 23 November 2009.

'Boimela O Nari-samabesh: the First National Feminist Book Fair-1999', http://membres.lycos.fr/ubinig/eventboimela.htm, last accessed 15 November 2009.

'Brown Spot: When and Where it Occurs', International Rice Research Institute, Manila, http://www.knowledgebank.irri.org/ricedoctor/index.php?option=com_content&view=article&id=557&Itemid=2762, last accessed 19 December 2009.

Chowdhury, Uttam Kumar et al., 'Groundwater Arsenic Contamination in Bangladesh and West Bengal, India', *Environmental Health Perspective*

(May 2000), http://findarticles.com/p/articles/mi_m0CYP/is_5_108/ai_633220
36?tag=artBody;col1, last accessed 15 November 2009.

'Factsheets on Chemical and Biological Warfare Agents', http://www.cbwinfo.
com/Biological/Pathogens/VC.html, last accessed 15 November 2009.

'Rabbits of Australia', http://web.archive.org/web/200001-200502re_/http://
rubens.anu.edu.au/student.projects/rabbits/history.html, last accessed
11 November 2009.

Wilson, Liz and Brant Wilson, 'Welcome to the Himalayan Orogeny' in http://
www.geo.arizona.edu/geo5xx/geo527/Himalayas/, last accessed 17 December
2009.

Index